Proceedings of the Society of Photo-Optical Instrumentation Engineers

Volume 78

Low Light Level Devices

For Science and Technology

March 22-23, 1976 ○ Reston, Virginia

Charles Freeman
Editor

Presented By
The Society of Photo-Optical Instrumentation Engineers
and the Society of Photographic Scientists and Engineers

In Cooperation With
Optical Sciences Center, University of Arizona
National Oceanic & Atmospheric Administration, U. S. Department of Commerce
U. S. Army Electronics Command, Night Vision Laboratory
Energy Research & Development Administration
Institute of Optics, University of Rochester
National Aeronautics & Space Administration, Ames Research Center
Department of the Navy, Office of Naval Research

ISBN 0-89252-105-8

Copyright © 1976 by the Society of Photo-Optical Instrumentation Engineers, 338 Tejon Place, Palos Verdes Estates, California 90274 USA. All rights reserved. This book or any part thereof must not be reproduced in any form without the written permission of the publisher. Printed in the United States of America.

LOW LIGHT LEVEL DEVICES FOR SCIENCE AND TECHNOLOGY

Volume 78

Contents

Introduction..v
Seminar Committee..vi

SESSION 1. GENERAL APPLICATIONS OF LOW LIGHT LEVEL DEVICES 1

Low-Light-Level Performance Analysis for Charge-Coupled Device TV Cameras 2
    Kenneth A. Hoagland, Fairchild Imaging Systems

Operation of CCDs in the Electron Bombarded Mode............................10
    L. Caldwell, J. Boyle, Night Vision Laboratory

Comparison of TV Imagers for Use in Low-Light-Level Imaging by Electron Beam Scan vs Solid-State Readout...14
    James A. Hall, Westinghouse Electric Corporation

Pyroelectric Vidicon Thermal Imager...23
    E. H. Stupp, Philips Laboratories

General Application of Microchannel Image Inverters.........................28
    Jon Tegethoff, Ferd Fender, Ni-Tec, Inc.

SESSION 2. APPLICATIONS OF LOW LIGHT LEVEL DEVICES IN PUBLIC SERVICE 33

Detection of Mine Hazards with Infrared Imagers.............................34
    Raymond M. Stateham, U. S. Bureau of Mines

Infrared is Not a Panacea—Use Discretion....................................41
    Alan J. Van den Berg, U.S. Army Corps of Engineers

The Technology behind X-Ray Security Systems................................44
    David J. Haas, Philips Electronic Instruments

Use of Night Vision Systems by the Land Manager.............................48
    Herbert J. Shields, USDA Forest Service

Unusual Applications of Image Intensification Devices.......................55
    Thomas M. Brennan, William H. Dyer, Baird-Atomic, Inc.

⇒⇒

SESSION 3. APPLICATIONS OF LOW LIGHT LEVEL DEVICES IN ASTRONOMY 59

A Large Imaging Array CCD Program 60
    Fred E. Vescelus, Jet Propulsion Laboratory; Gault A. Antcliffe, Texas Instruments

Astronomical Applications of Charge Injection Devices 65
    R. S. Aikens, C. R. Lynds, R. E. Nelson, Kitt Peak National Observatory

Intensified Charge Coupled Devices for Ultra Low Light Level Imaging 73
    Stanley Sobieski, NASA, Goddard Space Flight Center

Test Results on Intensified Charge Coupled Devices 78
    Jack T. Williams, NASA, Goddard Space Flight Center

A Photon Counting Array Photometer 83
    Douglas G. Currie, University of Maryland; John P. Choisser, Science Applications, Inc.

Low Light Level Imaging Devices for the Middle Ultraviolet 95
    G. Carruthers, J. Kervitsky, G. Hicks, C. Opal, Naval Research Laboratory

SESSION 4. APPLICATIONS OF LOW LIGHT LEVEL DEVICES IN MEDICINE 101

High-Resolution Low-Light-Level Video Systems for Diagnostic Radiology 102
    H. Roehrig, M. Frost, R. Baker, S. Nudelman, P. Capp, University of Arizona

Isocon Imaging for X-Ray Diagnostics 108
    Donald Sashin, David Gur, Clive W. Morris, John L. Ricci, School of Medicine, University of Pittsburgh

Image Intensifier Scintillation Cameras for Nuclear Medicine Applications 113
    Gerd Muehllehner, Searle Radiographics Inc. and Searle Analytic Inc.

Methods and New Approaches to the Calculation of Physiological Parameters by Videodensitometry .. 118
    Dan Kedem, Drora Kedem, D. P. Lindstrom, T. C. Rhea, Jr., J. H. Nelson, R. R. Price,
    C. W. Smith, T. P. Graham, Jr., A. B. Brill, Vanderbilt University Hospital

Results of Ratio Temperature Thermography 126
    E. L. Dereniak, H. Roehrig, W. L. Wolfe, Optical Sciences Center, University of Arizona

Radiometric FLIR for Thermography 131
    Scott P. Way, Honeywell Radiation Center

A Night Vision Aid as a Consumer Product 137
    James H. Burbo, ITT Electro Optical Products Division

Photoelectron Microscopy of Biological Surfaces—Excitation Source Brightness Requirements 143
    Rudy J. Dam, O. Hayes Griffith, University of Oregon

The Application of Low Light Level Video Techniques to Biomedical Research 155
    James A. Dvorak, William H. Schuette, Willard C. Whitehouse, National Institutes of Health

Author Index 161
Subject Index 162

LOW LIGHT LEVEL DEVICES FOR SCIENCE AND TECHNOLOGY

Volume 78

Introduction

Continuing improvement in performance and reduction in cost of image intensifier and low light level television have spurred an increasing number of applications of these and related technologies in the nonmilitary market. A similar process can be expected soon as technology for channel plate intensifiers, charge coupled devices, and thermal imagers matures sufficiently for the civilian market.

Seminar Committee

LOW LIGHT LEVEL DEVICES FOR SCIENCE AND TECHNOLOGY

Volume 78

General Chairman
Charles Freeman
Night Vision Laboratory, U. S. Army, Electronics Command

Co-Chairman
Carl M. Thomas
Strategic Technology Office, Defense Advanced Research Projects Agency

Chairman Session 1—General Applications of Low Light Level Devices
Ralph S. Levitt
Amperex Electronic Corporation

Chairman Session 2—Applications of Low Light Level Devices in Public Service
William Hawley
Night Vision Laboratory

Chairman Session 3—Applications of Low Light Level Devices in Astronomy
W. Kent Ford
Carnegie Institution of Washington

Chairman Session 4—Applications of Low Light Level Devices in Medicine
Sol Nudelman
University of Arizona

Session 1
GENERAL APPLICATIONS OF LOW LIGHT LEVEL DEVICES

Session Chairman
Ralph S. Levitt
Amperex Electronic Corporation

LOW-LIGHT-LEVEL PERFORMANCE ANALYSIS FOR CHARGE-COUPLED DEVICE TV CAMERAS

Kenneth A. Hoagland
Fairchild Imaging Systems
A Division of Fairchild Camera and Instrument Corporation
300 Robbins Lane, Syosset, New York 11791

Abstract

Charge-coupled device (CCD) area image sensors of the interline-transfer type include design features which enable the transfer and detection of signal charge packets of the order of tens of electrons. These sensors, which include devices with 190 x 244 and 380 x 488 elements, have application in all-solid-state TV cameras designed for use with very low levels of scene illumination. The designer of TV cameras which fully exploit this low-light performance potential is faced with a challenging constraint; extraneous noise effects must be minimized. In order to assess progress to date, both for sensor design and utilization techniques, the theoretical performance limitations imposed by non-cancellable noise of the detection process, and the effects of sensor CTF, are examined. This paper describes an analytical model for predicting the sensor-limited resolution-irradiance characteristics of CCD-TV camera systems as a function of responsivity, CTF, dark charge, on-chip amplifier NES, optical image contrast, and the required threshold signal-to-noise ratio. The results of the analysis are compared with recent experimental results approaching the predicted sensor-limited camera performance.

Introduction

Low-light-level television systems, utilizing high-gain image intensification principles, are capable of displaying output images as a pattern of scintillating bright spots corresponding to fluctuations in the input image signal. For a range of input irradiance and image contrast conditions, the observer's ability to perceive useful information is determined by the statistics of photon arrival "events" at the sensor input; i.e., the threshold performance of such systems can be described as photoelectron-limited. That such performance could ultimately be achieved was anticipated several decades ago, concurrent with the development of the first high sensitivity TV image sensor, the Image Orthicon camera tube. [1]

Although the progress of beam-scanned image sensor technology has been impressive, the minimum input format size and input irradiance levels required for useful system performance are constrained by the relatively low responsivity (typically 4 to 6 mA/W, 2854°K)* of photoemissive cathodes. It is this constraint which suggests the potential advantage of an all-solid-state image sensor; responsivities for silicon devices can be in the range of 25 to 100 mA/W, 2854°K.

If the threshold performance is to extend to very low irradiance levels, it is necessary to minimize extraneous noise sources, including dark current and detector/amplifier noise. Means for achieving the performance objectives have been implemented in the Fairchild developmental 190 x 244 and 380 x 488 CCD image sensors. [2] Charge transport in these devices is confined to a buried-channel region to prevent the interaction of signal electrons with surface state traps. Signals, in the form of individual charge packets, are detected by an on-chip low-noise distributed floating gate amplfier (DFGA) which has demonstrated an RMS noise equivalent signal (NES) of 10 to 20 electrons/pixel/frame. [3] That the necessary features can be combined to result in low light level imaging has been demonstrated; threshold resolution of test bar patterns at 1/2 the horizontal Nyquist limit has been achieved with a sensor highlight irradiance $\cong 10^{-5}$ W6m^2 (2 x 10^{-5} fc, 2854°K). [4]

This paper is primarily concerned with the analysis of performance of all-solid-state image sensors in TV camera systems, however, the results have broader application. The analysis is based on the postulates and experimental results of Rose, Coltman, and others, defining the observer's perception of signal and noise effects for specific types of displayed images. The general approach employed is similar to previously described analyses for intensifier-aided photo-optical systems [5] and beam-scanned image sensors. [6]

Signal-to-Noise Ratio Considerations

The observer threshold signal-to-noise ratio (SNR_o) required for the perception of information from both "live" and hard-copy displays has been extensively investigated over the past few decades. Although psychophysical aspects of the perception process have yet to be established on a firm theoretical base, a number of investigators have demonstrated the validity of system analysis techniques formulated to conform with the results of controlled experiments defining SNR_o. Results specifically applicable to this camera system analysis are shown in Table I.

The Table I results suggest that the eye-brain system is very effective in extracting information when signals are masked by temporal noise. This hypothesis is supported by an experiment conducted by Mirkin and Campana (NADC, Warminster, Pa.). In this experiment, a video recorder was used to record single TV frames, which were subsequently played back for comparison with the normal "live" viewing condition of the same test pattern information. It was found that the signal-to-noise ratio for the single stored-frame case had to increase by a factor of 5 db for equivalent threshold resolution of multiple-bar test patterns, corresponding to an increase in SNR_o by a factor of approximately 1.8X.

* As determined with a calibrated tungsten lamp source at the color temperature specified.

TABLE I. THRESHOLD SNR_d VALUES FOR IMAGE PERCEPTION

IMAGE TYPE	SNR_d THRESHOLD, 50% PROBABILITY WHEN VIEWED WITH OPTIMUM VIEWING MAGNIFICATION OR DISTANCE	
	STORED IMAGE	LIVE IMAGE
APERIODIC DISKS, SQUARES, OR RECTANGLES	$\cong$ 5/1 (a), (b)	$\cong$ 3/1 (c)
THREE BAR TEST PATTERN (5:1 HEIGHT-TO-WIDTH RATIO) WITH SNR_d COMPUTED FOR A SINGLE BAR AREA	$\cong$ 3.6/1 (d)	$\cong$ 2/1 (d) 2.6 to 1.5 (c) (decreases for smaller bars)

REFERENCES:
(a) Rose, A., "The Sensitivity of the Human Eye on an Absolute Scale", J. Opt. Soc. Am. 38, 196–208, 1948.
(b) Chambers, R.P. and Courtney-Pratt, J.S., Photo. Sci. Engng., 13, 286–298, 1969.
(c) Rosell, F.A. and Willson, R.H., In *Perception of Displayed Information,* Chapter 5, L.M. Biberman, ed. Plenum, New York, 1973.
(d) Schade, O.H., Sr. "The Resolving Power Functions and Quantum Processes of Television Cameras", RCA Rev. 28:3, 460–535, 1967.

The Table I results were obtained with simulated near-ideal test systems, or were corrected where necessary to include the effects of response and noise degradations prior to the observer input. In addition, the observers were allowed to adjust viewing conditions to optimize pattern detection. Thus, SNR_o, the signal-to-noise ratio perceived by the observer is assumed to be equivalent to SNR_d, the measured value of SNR referenced to the display output. Under these conditions, the signal perceived by the observer, S_o, is given by:

$$S_o = \bar{d}_w - \bar{d}_b \qquad (1)$$

where $\bar{d}_w$ and $\bar{d}_b$ are event densities corresponding to the highlight (subscript w) and low-light (subscript b) regions of the display equal in area to the area, a_o, of the aperiodic or test bar image. Bar superscripts denote average values observed for a large number of eye integration time intervals, t_e. Defining $\bar{d}_{cw}$ and $\bar{d}_{cb}$ as the average event densities in a unit cell of area $a_c < a_o$, as determined during a time interval $t_f < t_e$, yields:

$$S_o = (\bar{d}_{cw} - \bar{d}_{cb})(a_o / a_c)(t_e / t_f) \qquad (2)$$

This form of the S_o expression is useful for CCD-TV analysis where the displayed image is made up of discrete element areas corresponding to unit cells of area a_c at the image sensor format, and where t_f is the signal integration or frame time for the TV system. For the stored image case (or for live viewing with image intensifiers) $t_e / t_f = 1$, hence, these time-dependent terms may be deleted.

Observer integration effects for noise are based on the postulate that the RMS noise perceived from an elemental area is determined by $(\bar{d})^{1/2}$ where d denotes the average event density in the area during the observation interval. For the stored image case considered by Rose[7], the observed RMS noise N_o was determined by $(\bar{d}_w)^{1/2}$ i.e., by the average event density in highlight regions. For the live image case, Coltman[8] postulated that the eye perceives noise from equal area regions of highlight and background yielding an RMS noise term of the form $N_o = (\bar{d}_w + \bar{d}_b)^{1/2}$. Although the Coltman and Rose postulates differ, the effect on predicted results can be relatively minor, as shown by Legault[9].

Thus, for the analysis which follows, N_o is defined as:

$$N_o = \left[(\bar{d}_{cw} + \bar{d}_{cb})(a_o / a_c) t_e / t_f\right]^{1/2} \qquad (3)$$

Hence, from equations (2) and (3):

$$SNR_o = S_o/N_o = \frac{(\bar{d}_{cw} - \bar{d}_{cb})(a_o / a_c)^{1/2}}{(\bar{d}_{cw} + \bar{d}_{cb})^{1/2}} (t_e / t_f)^{1/2} \qquad (4)$$

The application of (4) requires a known value for t_e. Results reported by Schade[10] indicate t_e values in the range 0.2 to .05 second, dependent on display luminance. The frequently assumed value of 0.2 second appears to be applicable only for unrealistically low luminance conditions near the vision threshold. The Mirkin and Campana experiment, and the ratio of SNR_d values for live vs stored images, suggest the value $t_e \cong 0.1$ second, i.e., the eye-brain system effectively integrates noise over approximately six TV fields (3 frames) for the conventional 30/second frame rate condition.

Camera System Analysis

When SNR_d values for the display output can de defined, the additional effort required to predict threshold resolution performance is a careful step-by-step analysis of the noise sources and the noise and signal transfer characteristics for each component of the system. The concept is simple; assuming an ideal linear system, without noise sources and without MTF degradation, the output signal will be a replica of the input and SNR_d is determined by the photogenerated signal and noise-in-signal characteristics at the system input. For actual systems, the effects of added noise and the transfer functions which operate on signal and noise must be included.

At least for CCD image sensors, it is convenient to define the sensor signal and noise characteristics with reference to the average number of carriers (i.e. $\bar{n}$ electrons) contained in the individual charge packets sensed by the output detector. Thus, for ideal CCD systems, equation (2) becomes:

$$\text{Observed signal, } S_o = (\bar{n}_w - \bar{n}_b)(a_o/a_c)(t_e/t_f) \tag{5}$$

where: $\bar{n}_w$ and $\bar{n}_b$ are the number of electrons/pixel/frame for the highlight and low-light regions, respectively

a_o is the test object area at the input format

a_c is the sensor unit cell (pixel) area

t_e is the eye integration time, seconds

t_f is the frame time, seconds

If signal and background shot noise and amplifier noise are the dominant components of system noise, the observed RMS noise, N_o, is given by the RSS summation of noise effects for both highlight and low-light regions:

$$\text{Observed noise, } N_o = (\tilde{n}_{nw}^2 + \tilde{n}_{nb}^2)^{1/2} (a_o/a_c)^{1/2} (t_e/t_f)^{1/2}$$
$$= (n_w + n_b + 2\tilde{n}_{an}^2)^{1/2} (a_o/a_c)^{1/2} (t_e/t_f)^{1/2} \tag{6}$$

where: $\tilde{n}_{nw}$ and $\tilde{n}_{nb}$ are the RMS noise contributions for highlight and low-light regions, respectively, expressed as equivalent electrons/pixel/frame.

$\tilde{n}_{an}$ is the amplifier NES, electrons/pixel/frame.

Defining the CCD electron image contrast, $C_q = (\bar{n}_w - \bar{n}_b)/\bar{n}_w$ and substituting this expression in (5) and (6) yields:

$$S_o = C_q \bar{n}_w (a_o/a_c)(t_e/t_f) \tag{7}$$

and

$$N_o = \left[\bar{n}_w(2-C_q) + 2\tilde{n}_{an}^2\right]^{1/2} (a_o/a_c)^{1/2} (t_e/t_f)^{1/2} \tag{8}$$

For the test object to be observed, the threshold SNR_o must be equal to or greater than $K_{s/n}$, an empirically determined constant for a particular test pattern geometry and detection probability condition (ref. Table I); thus:

$$SNR_o = S_o/N_o \geq K_{s/n}$$
$$= C_q \bar{n}_w (a_o/a_c)^{1/2} (t_e/t_f)^{1/2} / \left[\bar{n}_w(2-C_q) + 2\tilde{n}_{an}^2\right]^{1/2} \tag{9}$$

which is equivalent to:

$$n_w = K_{s/n} \left[n_w(2-C_q) + 2\tilde{n}_{an}^2\right]^{1/2} (a_o/a_c)^{1/2} (t_f/t_e)^{1/2} /C_q \tag{10}$$

A quadratic solution for $\bar{n}_w$ can be obtained from (10), but a more useful form of the solution is obtained if the parameters defining $\bar{n}_w$ and C_q in actual (non-ideal) camera systems are introduced. For the determination of sensor-limited system performance, these parameters are:

$\bar{n}_{sw}$ the highlight signal measured with reference to the average sensor dark level, n_d, at image spatial frequency $f \rightarrow 0$

C, the input optical image contrast, defined as $C = (\bar{n}_{sw} - \bar{n}_{sb})/\bar{n}_{sw}$,

$\bar{n}_d$, the average level due to thermally generated dark charge (as determined without input irradiance)

M_c, the contrast transfer function for the image defined at optimum phasing for the spatial frequency range from 0 to f_n, the Nyquist limit,

Figure 1 illustrates the functional relationships for these parameters. It should be noted that the amplitude levels denote average levels of electrons/pixel/frame, corresponding to measurements made at the sensor video output.

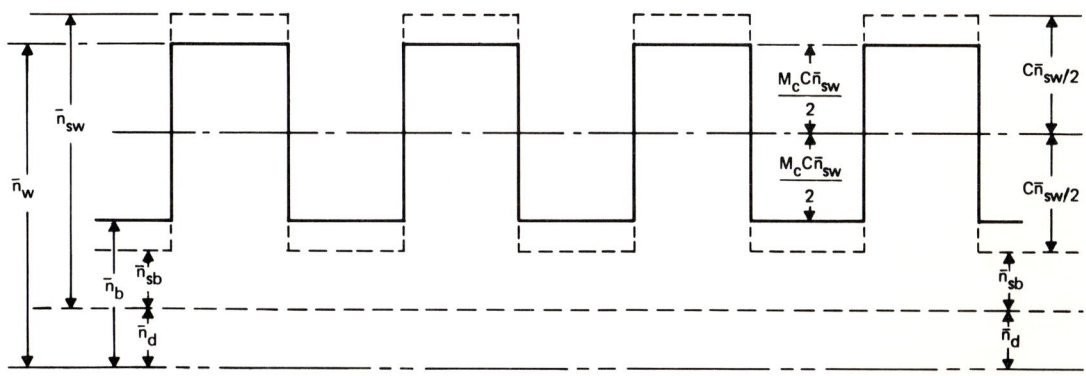

OPTICAL IMAGE CONTRAST, C

$$C = \frac{\bar{n}_{sw} - \bar{n}_{sb}}{\bar{n}_{sw}}$$

ELECTRON IMAGE CONTRAST, C_q

$$C_q = \frac{\bar{n}_w - \bar{n}_b}{\bar{n}_w} = \frac{M_c C \bar{n}_{sw}}{\bar{n}_w}$$

where $\bar{n}_w = \bar{n}_d + \bar{n}_{sw} - \frac{C\bar{n}_{sw}}{2} + \frac{M_c C \bar{n}_{sw}}{2} = \frac{\bar{n}_{sw}}{2}(2 - C + CM_c) + \bar{n}_d$

M_c IS CCAID CONTRAST TRANSFER FUNCTION, $0 < M_c < 1$

FIGURE 1. CONTRAST AND CTF PARAMETERS

The quadratic solution of (10) with the defined parameters included is:

$$\bar{n}_{sw} = \frac{K_{s/n}^2 a_k (2-C) + K_{s/n} a_k^{1/2} \left[K_{s/n}^2 (2-C)/4 + 2M_c C^2 (\bar{n}_d + \tilde{n}_{an}^2) \right]^{1/2}}{M_c^2 C^2} \qquad (11)$$

where $a_k = (a_c/a_o)(t_f/t_e)$

$= (a_c/a_o)/3$ for $t_f = 1/30$ and $t_e = 1/10$ seconds

If the image sensor has significant amplifier noise ($\tilde{n}_{an} \geq 10$) the terms containing (2-C) are relatively small and may be deleted. Regrouping the remaining terms yields the approximate solution:

$$\tilde{n}_{sw} \cong K_{s/n} \left[2a_k (\bar{n}_d + \tilde{n}_{an}^2) \right]^{1/2} / CM_c^{3/2} \qquad (12)$$

for $M_c C^2 (\bar{n}_d + \tilde{n}_{an}^2) > 1$

For camera systems with a noiseless form of image intensification prior to the readout sensor, the effect of pre-readout gain G is to decrease the significance of readout sensor noise by the factor 1/G. When G is sufficiently high, the threshold signal requirements will be a function of the signal and noise-in-signal statistics at the photoreceptor input and the system will be photoelectron limited. The solution of equation (11) for the photoelectron-limited case occurs when $(\bar{n}_d + \tilde{n}_{an}^2) \rightarrow 0$, thus:

$$\bar{n}_{sw} \cong K_{s/n}^2 \, a_k \, (2-C)/M_c^2 C^2 \tag{13}$$

$$\text{for } (\bar{n}_d + \tilde{n}_{an}^2) < 0.1$$

The data illustrated in Figure 2 was derived using (12) and (13). These data indicate that high-performance, low-noise, all-solid-state image sensors are potentially capable of exceeding the low-light-level performance of existing beam-scanned image sensors (such as the ISIT) if the comparison criteria includes equivalent format size and the detection of low contrast scene information.

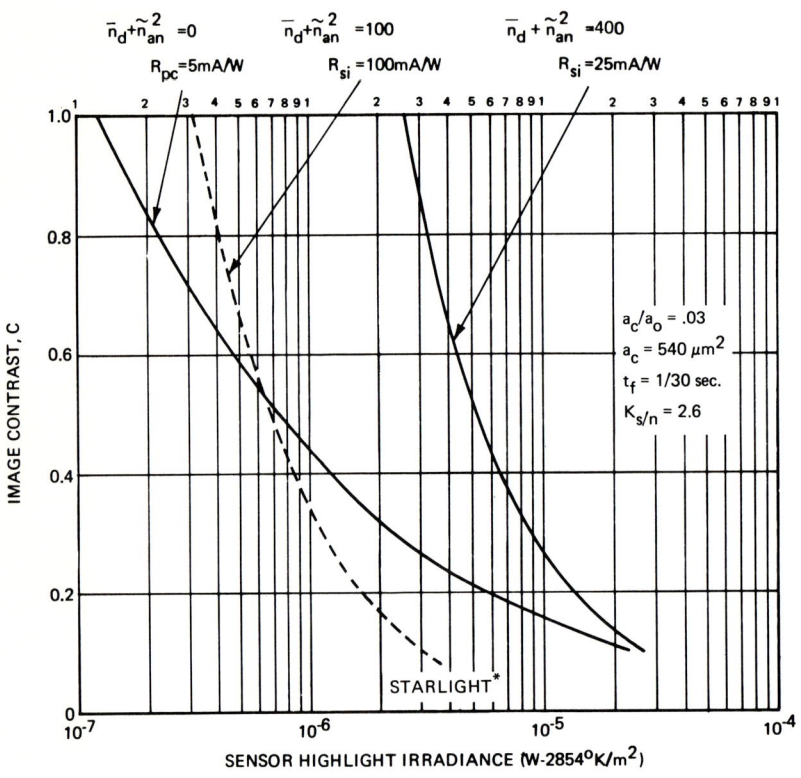

*FOR CLEAR NIGHT SKY CONDITIONS WITH
T1.2 OPTICS AND 50% HIGHLIGHTS REFLECTANCE

FIGURE 2. CONTRAST – IRRADIANCE CHARACTERISTICS AT 1/2 NYQUIST LIMIT, $M_c = 1.0$

Applications

The application of equation (11) can be illustrated using parameters appropriate to the Fairchild developmental type CCAID-244 and CCAID-488 charge coupled area imaging sensors. The unit cell (pixel) area for these sensors is given by $a_c = h \times w = 18 \, \mu m \times 30 \, \mu m$. The 30 μm width dimension includes a photosensor site and an adjacent light shielded area which is one half stage of a vertical two-phase charge transport register. Typical measured responsivities for prototype devices are in the range of 25 to 30 mA/W (2854°K). Assuming 25 mA/W, and 1/30 second frame (integration) time, an $\bar{n}_{sw}$ value of 10 electrons/pixel/frame is obtained with an irradiance at the sensor input of 3.6×10^{-6} W/m^2.

Figure 3 illustrates a test target designed to simplify correlation of predicted and experimental results. The target contains 3-bar groups of 5:1 height to width ratio, dimensioned to correspond 1/2 and 1/4 Nyquist-limit horizontal resolution for a 190 x 244 element array. When imaged at the proper magnification ratio, as determined by the format border markings, the test bar images can be precisely positioned for optimum spatial phasing.

Table 2 illustrates a typical $\bar{n}_{sw}$ computation with conditions corresponding to a previously reported laboratory test of a CCAID-244C image sensor at low-light-level. [3] During this test threshold imaging was demonstrated at 1/2 Nyquist limit horizontal resolution with an $\bar{n}_{sw}$ value determined to be at 25 $\bar{e}$/pixel/frame. (See Figure 4).

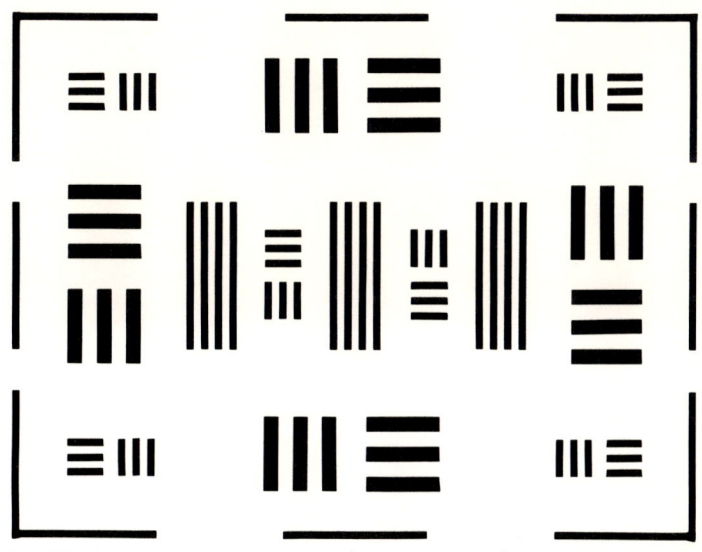

The vertical and horizontal bar groups correspond to 1/4 and 1/2 Nyquist-limit horizontal resolution

FIGURE 3. TEST TARGET FOR 190 X 244 ELEMENT CCD ARRAYS

TABLE 2. CCAID-244C THRESHOLD SIGNAL COMPUTATION

	CONDITIONS
From Equation (12) $\bar{n}_{sw} = K_{s/n} \left[2a_k (\bar{n}_d + \tilde{n}_{an}^2) \right]^{1/2} / CM_c^{3/2}$	
$a_k = a_2 = (a_c/a_o)/3 = .01$	1/2 Horizontal Nyquist limit; 30 F/S; $a_c = 18 \times 30$, $a_o = 60 \times 300$
$K_{s/n} = 2.0$	Table I, ref. (d)
$\bar{n}_{sw} \cong 2.0 \left[.02 (\bar{n}_d + \tilde{n}_{an}^2) \right]^{1/2} / M_c^{3/2}$	C = 1

$\bar{n}_d$	$\bar{n}_d + \tilde{n}_{an}^2$	M_c	$\bar{n}_{sw}$	
0	400	0.8	7.9	
300	700	0.8	10.5	DFGA NES = $\tilde{n}_{an} = 20\,\bar{e}$
600	1000	0.8	12.5	
1200	1600	0.8	15.9	

The sensor used for this experiment exhibited very low dark current density, estimated to be in the range from 4 to 10 nA/cm^2 at 25°C. The predicted $\bar{n}_{sw}$ values as a function of dark level n_d indicate relatively small effects are to be expected as a result of cooling below the level corresponding to $\bar{n}_d = \tilde{n}_{an}^2$, in accordance with the results observed during test runs covering a range from 0°C to -40°C.

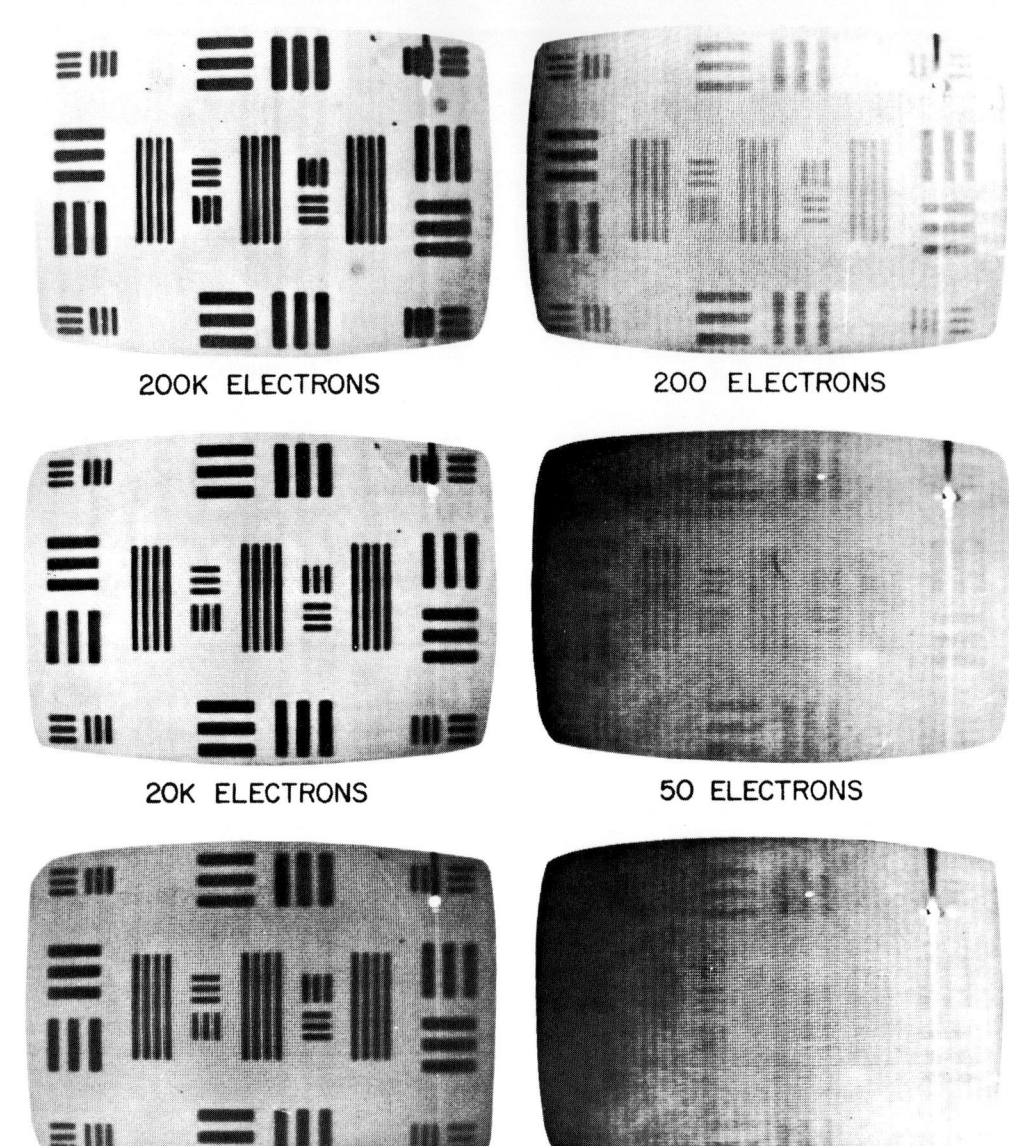

FIGURE 4. TEST BAR IMAGING WITH THE DFGA OUTPUT OF A 190 X 244 SENSOR. EXPOSURE TIME: 0.1 SECOND, TEMPERATURE: 0°C, 30 FRAMES/SECOND ELECTRON COUNTS REPRESENT ELECTRONS/PIXEL/FRAME IN IMAGE HIGHLIGHT REGIONS

Conclusions

An analytical method for predicting low light level performance thresholds for TV camera systems using CCD or other types of charge-transfer-device image sensors has been described. The analysis predicts threshold performance near that observed during an initial test of a cooled 190 x 244 element CCD specifically developed for low light level camera applications.

The analysis results in a quadratic solution for the required highlight signal $\bar{n}_{sw}$, as a function of a bar-element threshold resolution parameter (a_k), the optical image contrast C, and the sensor CTF, dark signal, and amplifier noise parameters. Special case solutions have been derived for comparing photoemissive-cathode intensified CCD or ISIT sensors with all-solid-state sensors of identical format size; a high responsivity all-solid-state sensor with NES = $10\bar{e}$ is predicted to exhibit superior performance when imaging low contrast scene information.

Acknowledgement

The development of low-light-level CCD image sensors for TV applications was in part supported by Naval Electronics Systems Command Contract No. 00039-73-C-0015, directed by L. Sumney. The contributions of many members of the Fairchild organization during the CCD development effort, particularly G. Amelio, H. Balapole, H. Dean, R. Dyck, J. Early, I. Hirschberg, C.K. Kim, W. Steffe, L. Walsh and D. Wen are gratefully acknowledged.

The author is also grateful to S. Campana of the Naval Air Development Center, Warminster, PA, for many helpful discussions concerning low light level TV system performance.

References

1. Rose, A., In *Vision, Human and Electronic,* Preface; Chapter 3, Plenum, New York, 1973

2. Steffe, W., Walsh, L., and Kim, C.K., "A High Performance 190 x 244 CCD Area Image Sensor Array", Proc. of the Int'l. Conf. on the Application of Charge-Coupled Devices, pp 101–108, San Diego, CA, Oct. 1975.

3. Wen, D.D., "Low Light Level Performance of CCD Image Sensors", loc. cit., pp 109–119.

4. Hoagland, K.A. and Balopole, H.L., "CCD-TV Cameras Utilizing Interline-Transfer Area Image Sensors", loc. cit., pp 173–180.

5. Hoagland, K.A., "Design Techniques for Low Light Photo-Optical Equipment", Proc. of the 13th Annual SPIE Technical Symposium, pp 89–93, Washington, D.C., Aug. 1968.

6. Rosell, F.A., In *Photoelectronic Imaging Devices,* Vol. 1, Chapter 14; Vol. 2, Chapter 22, Plenum, New York, 1971.

7. See Table I, ref. (a).

8. Coltman, J.W., "Scintillation Limitations to Resolving Power in Imaging Devices", J. Opt. Soc. Am. 44:3, 234–237, 1954.

9. Legault, R.L., In *Photoelectronic Imaging Devices,* Vol. 1, Chapter 4, Plenum, New York, 1971.

10. Schade, O.H., Sr., "Optical and Photoelectric Analog of the Eye", J. Opt. Soc. Am. 46:9, 721–239, 1956.

OPERATION OF CCD's IN THE ELECTRON BOMBARDED MODE

L. Caldwell and J. Boyle
Night Vision Laboratory
Fort Belvoir, Virginia 22060

Abstract

To fully realize the potential of charge coupled devices for solid state low light level imaging additional gain before the array appears necessary. This can be accomplished either by coupling the device to an image intensifier tube optically or by operating in the electron bombarded mode (i.e. EBS). As in the case of the electron bombarded silicon diode array vidicon the performance in the electron-in mode is expected to be superior to the performance in the optically coupled mode.

100 x 160 element thinned CCD arrays have been operated in the EBS mode and the results indicate that substantial improvement in low light level performance can be realized when compared to direct photon-in performance. Imaging has been demonstrated at an equivalent signal level of 5 electrons/pixel (i.e. Actual No./Gain).

Introduction

Two basic approaches to solid state low light level imaging are presently under investigation. The most thoroughly investigated up to the present time involves the direct photon-in mode, in which the highest quality arrays are used, cooling to about $-20°C$ is required, and sophisticated low noise detection schemes are employed. Even so, results to date, both analytical[1] and experimental,[2] indicate that for low light level imaging corresponding to starlight conditions, (i.e. scene illuminations of $10^{-4} - 10^{-5}$ ft -cd) additional gain before the CCD chip is necessary.

Gain can be achieved in the most efficient way be operating the array in the electron bombarded mode as shown in Fig. 1.

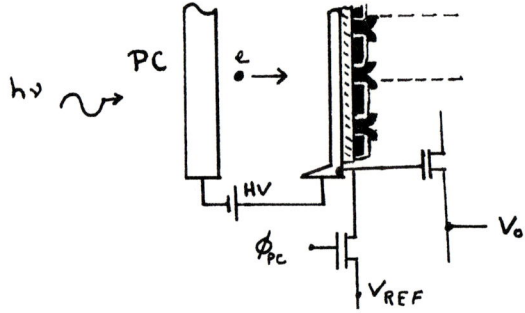

Fig. 1 Schematic Diagram of EBS-CCD

This is closely analogous to the electron bombarded silicon diode array vidicon and much of the technology developed for the target fabrication in that device is applicable to the EBS-CCD. The chips must be thinned to approximately 1/2 mil and the back surface accumulated for optimum electron gain. Eventually it's expected that electron gains of 2000 at 10 kilovolts will be routinely achieved, however, at the present time due to processing limitations, (primarily temperature restrictions) gains of 1000 at 10 kilovolts are more typical (on a recent array a gain of 2000 at 10 kilovolts was observed).

In addition to low light level operation, EBS CCD's offer the advantage of room temperature operation, spectral selectivity of the photocathode, and when operated in the photon-in mode, thinned arrays can be fabricated with excellent blue response (i.e. QE = 70% at 4000A°) and white light responsivity (i.e. 100 ma/watt for 2854°K tungsten).

In this paper initial results with electron-in imaging are presented and compared with photon-in results. While the photon-in results are rather typical of those obtained on area arrays using simple precharge amplifiers,[3] and are several orders of magnitude away from photon noise limited performance, substantial low light level performance improvements can be realized when these arrays are operated in the EBS mode. With low gain arrays (i.e. ~600 at 10kV) electron imaging has been demonstrated at an equivalent signal level of 5 electrons/pixel. (i.e. Actual signal/Gain)

Description of the Image Sensor

The image sensors used in this study were supplied by Texas Instruments under Contract DAAG53-75-C-0191. They are area imaging arrays containing 100 x 160 resolution elements (.9 x .9 mils²). The arrays are buried channel, 3 phase, thinned, backside illuminated structures and have been described in some detail previously.[4] The electrode structure is double level aluminum seperated by a thin (~2500A°) layer of anodized aluminum oxide. It can be seen (Fig. 1) that equivalent phases on adjacent pixels are provided by different levels of metallization. This can lead to a fixed pattern type of interference at low light level (i.e. "line pairing") as will be seen. Readout is accomplished through a simple precharge transistor and source follower. The load for the source follower is a 4.7k resistor located off chip.

All data were taken at room temperature with the array's operated in the full frame mode and shuttered during readout. The frame time was 36 msec and the readout rate was 1 MHz. The spectral response of three representative arrays is shown in Fig. 2.

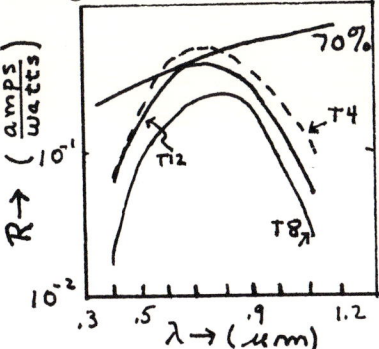

Fig. 2 Spectral Response of CCD Imaging Arrays

While chips with higher blue response (and therefore higher electron gain) have been reported[4] these were the best available. Additional data of these arrays (labeled T4, T8 and T12) is shown in Table I.

Table I

Array	R $\left(\frac{ma}{watt}\right)$	CTE	QE (4000A°)	Gain (10kV)	$J_D \left(\frac{na}{cm^2}\right)$	T (microns)
T4	~100	.9992	5%	630	13	13
T8	73	.998	15%	-	31	8
T12	91	.9997	14%	600	8	8

The thickness of the arrays has been estimated from the measured photoresponse at 1.0 micron assuming that the front surface metallization is 100% reflective. The responsivity in Table I is that measured with tungsten illumination at 2854°K and J_D is the dark current of the array at 24°C.

Low Light Level Performance

Photon-In

Imaging of a 100% contrast Air Force resolution pattern is shown in Fig. 3. The high light level picture (i.e. 35 x 10^5 electrons) is taken at about 40% full well. Full well is estimated to be 8.75 x 10^5 electrons/pixel, measured from the onset of saturation of the precharge current.

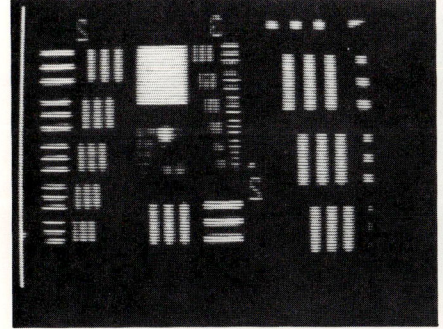
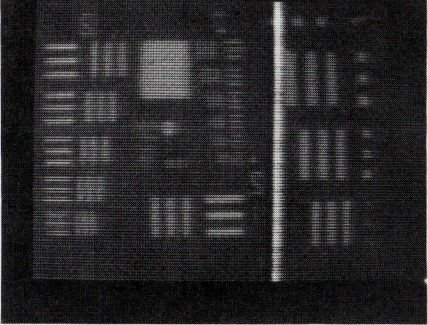
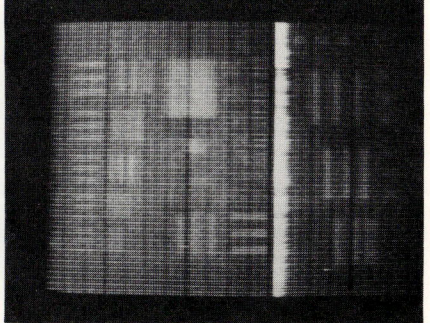

A 350,000 $\frac{electrons}{pixel}$ B 10,500 $\frac{electrons}{pixel}$ C 2,400 $\frac{electrons}{pixel}$

Fig. 3 Photon Imaging on array T12 at various light levels. Exposure time was .2 sec.

Substantial differences can be noticed between the vertical and horizontal resolution. This can be caused by CTE degradation, however, in the present case the CTE is high (i.e. .9997 from Table I) and the degradation is independent of position on the array. The degradation is believed to be in the video amplifier chain.

Several defects can be noticed as the light level is decreased. A blooming line defect is present, which is not visible under highlight conditions. The apparent dark line defects are not due to the array but are caused by the video electronics. Line pairing (i.e. alternate light and dark horizontal lines) can be seen in the image at 2400 electrons/pixel. A more severe case of the effect is shown in Fig. 4 for sensor T8. Line pairing is caused by slight differences in cell geometry between cells formed by upper layer and lower layer metallizations. Minor chip redesign should remove this problem.

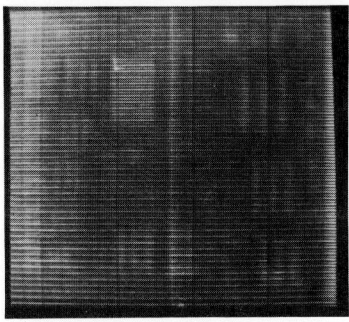

Fig. 4 Imaging at 2400 electrons. Array T8

Fig. 5 Limiting Resolution vs. Light Level. Photon-In.

Limiting resolution vs. light level is shown in Fig. 5 for arrays T8 and T12. The degradation in horizontal resolution noted on the photographs is obvious here. The highlight irradiance is expressed in terms of electrons/pixel. This can be converted to the more conventional irradiance in watts/m² by the expression:

$$I \; \frac{watts}{m^2} = \frac{eN}{R \tau_I A_p} \qquad (1.)$$

where N is no electrons/pixel, R is the chip responsivity (or photocathode responsivity if operated in the EBS mode, τ_I is the chip integration time), (18 msec for all data presented here), and A_p is the area of a single pixel.

Electron-In

The electron bombarded results were obtained on a modified RCA (C-33052) 60-18 zoom image intensifier. A modified flange with the CCD array mounted on a tube compatible header was inserted in place of the phosphor. Imaging was done with a UV source and a gold cathode.

The electron image quality is shown in Fig. 6 for sensor T4. The low light level picture corresponds to 3200 electrons per well, but with a gain of 630 at 10kV. this is equivalent to imaging 5 electrons per well incident on the array.

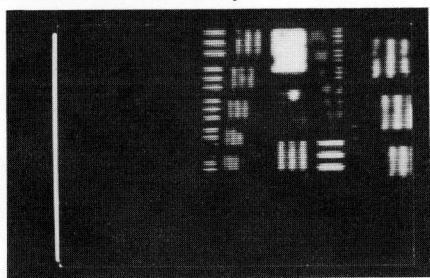 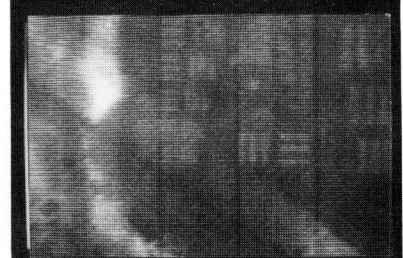

A $\frac{300,000}{630}$ = 480 electrons   B $\frac{3,200}{630}$ = 5 electrons

Fig. 6 Electron Imaging vs. Light level. Array T4

Similar imaging for array T12 is shown in Fig. 7. It can be seen that the line pairing is more severe then in array T4 and that the resolution is significantly degraded.

In addition a dark area of shading exists in the upper right corner characteristic of a region of high back surface recombination velocity. The reason for the degraded resolution of sensor T12 is not immediately apparent. Array T12 has a higher charge transfer efficiency (see Table I), and estimates of the thickness using the response at 1.0 micron indicate that array T12 is thinner than array T4 (i.e. 8 vs. 13 microns). Thus, both transfer efficiency and electron diffusion effects indicate that array T12 would have better MTF.

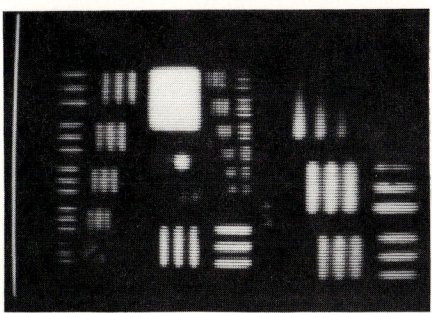
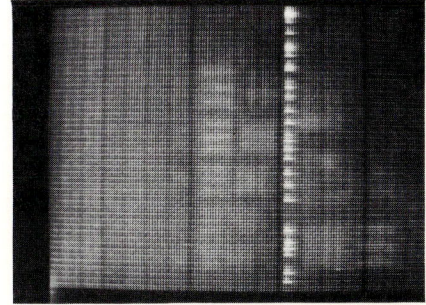

A. 8 electrons B. 80 electrons

Fig. 7 Electron Imaging vs. Light Level. Array T12

The low light level limiting resolution for these arrays operated in the electron bombarded mode is shown in Fig. 8.

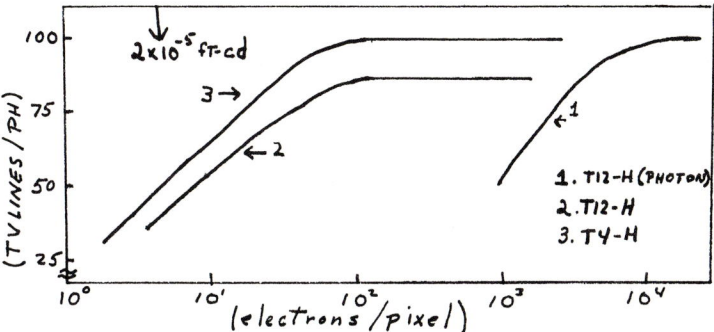

Fig. 8 Limiting Resolution vs. Light level. Electron-In

The performance advantage of the electron bombarded mode can be clearly seen. In order to obtain a more accurate measure of the performance improvement one, of course, needs to consider the responsivity of the photocathode used compared to the responsivity of the chip. One can see from Table I that compared to a typical 7 ma/watt S-20 photocathode, the chip responsivity can be higher by a factor of 15. Even so the EBS mode has demonstrated improved performance. On the other hand, when comparing to front side illuminated CCD's where 35 ma/watt is a more typical responsivity, the difference in responsivities is significantly less. Also shown in Fig. 8 is the sensor irradiance corresponding to clear moonless night sky conditions for a 7 ma/watt photocathode. (i.e. 2×10^{-5} ft-cd on the cathode).

Conclusions

It has been demonstrated that CCD's can be operated in the electron bombarded mode. Further, even arrays with low gain, operated with video channels not necessarily designed for low light level applications have demonstrated imaging better than the best cooled photon-in results. It remains, however, to demonstrate high resolution EBS performance in order to fully realize the potential of the EBS-CCD.

Acknowledgement

The authors would like to express their gratitude to Mr. Wayne Tardiff and Mr. Samuel Tomarchio for their expert technical help and Dr. Ed Smith for many helpful discussions.

References

1. Barton, J. B., Cuny, J. J. and Collins, D. R., "Performance Analysis of EBS-CCD Imaging Tubes/Status of ICCD Development," Proceedings of International Conference on Charge Coupled Devices, San Diego, California, Oct. 1975.

2. Wen, D. D., "Low Light Level Performance of CCD Image Sensors," Proceeding of International Conference on Charge Coupled Devices, San Diego, California, Oct. 1975.

3. Dyck, R. H. and Jack, M.D., "Low Light Level Performance of a Charge-Coupled Area Imaging Device," Proceedings of International Conference on Technology and Applications of Charge Coupled Devices, Edinburgh, Scotland, Sept. 1974.

4. Antcliffe, G. A., et. al. "Large-Area CCD Imagers for Spacecraft Applications," Proceedings of the Symposium on Charge Coupled Device Technology for Scientific Imaging Applications" Jet Propulsion Laboratory, Pasadena, California, March 1975.

COMPARISON OF TV IMAGERS FOR USE IN LOW-LIGHT-LEVEL IMAGING
BY ELECTRON BEAM SCAN VS. SOLID-STATE READOUT

James A. Hall
Westinghouse Electric Corporation
Advanced Technology Laboratory
Baltimore, Maryland 21203 USA

Abstract

Solid-state imagers can displace electronic beam scanned imagers primarily in applications which conventional TV has not satisfied. Included are imaging from a moving platform like a weather satellite, imaging with very large intrascene dynamic range, imaging of very low contrast scenes including imaging in the infared where solid-state arrays in optomechanical scanners provide basically better performance and TDI techniques are increasing their advantage, and very low light imaging where "photon counting" techniques using an ICCD with an external computer memory can eliminate problems of dark current and of preamplifier noise in long time exposures and provide higher signal-to-noise ratio for improved radiometric accuracy. In direct comparison, solid-state imagers lead for larger effective dynamic range, greater maximum signal-to-noise ratio, freedom from lag, geometric fidelity and metricity, and stability of characteristics, as well as the obvious compactness, long life, and low operating voltages. Tubes lead for the greatest number of resolvable elements per frame, for producing images with better element-to-element uniformity, and for maximum output data rate on a single terminal.

Introduction

Any comparison of solid-state and electron beam scanned imagers should show which applications can best be satisfied by each candidate. We consider first three applications which tubes do not handle well, where solid-state imagers have real advantages. Then the characteristics of tubes and solid-state imagers are compared directly as a guide to sensor selection for other applications. At this time, solid-state imagers seem unlikely to displace camera tubes for broadcast TV-type applications but have potential advantages for a number of scientific or military applications, especially those for which the traditional 4 x 3 aspect ratio image of a mostly stationary scene is not appropriate.

Meteorological Satellites

In early weather satellites vidicon-type television cameras transmitted a series of "snapshots" of the earth's surface and its cloud cover. Image motion was compensated mechanically during each exposure to avoid blurring. Each frame was printed separately, and the observer had to cut and paste to provide area coverage beyond the 300 to 500 element square raster of each frame. In contrast, the U.S. Defense Meteorological Satellite uses a silicon detector in each of two mechanical scanners to provide simultaneous 2 mile resolution and 1/3 mile resolution imaging. Cross-track scan is provided by rotating mirrors while vehicle motion provides along-track scan. As shown in Fig. 1, two satellites operate in nearly polar orbits, with one vehicle crossing the equator from south to north at local noon, and the other at about 7:30 in the morning. The output data is the image of a continuous strip 1,655 nautical miles wide and is printed on long strips of 9-1/2 inch film, each containing the record of three orbits.

The ground irradiance for the noon-midnight sensor varies from a maximum at the noon south-to-north crossing of the equator, to a minimum at the midnight north to south crossing, over at least $10^8:1$. The silicon photoconductive sensor is linear over this range, and output signal level is maintained relatively constant by use of a light sensor actuating an amplifier gain control. For sun elevation less than 6 degrees, the sun elevation angle is measured instead and the gain control adjusted from computed irradiance values. The latter scheme is used exclusively on the early morning satellite, which must compensate for scene irradiance variations along a single scanning line of up to $3 \times 10^5:1$ when the scan line spans the terminator. Scene reflectance variations which constitute the information add one to two orders of magnitude to the range of irradiances at the sensor. To use a conventional TV sensor for the early morning satellite would thus require accommodating a dynamic range of $10^7:1$ in a single scene, a manifestly impossible task.

Samples of imagery obtained are given in Fig. 2 and 3. Fig. 2 was taken over Italy by the early morning satellite at 2-nmi resolution. The crossed grid lines occur as the amplifier gain is changed along each scanning line and along track. The image is more than adequate to record the bright cloud cover to the east over Yugoslavia and Albania where the sun is higher and the absence of clouds over Sardinia,

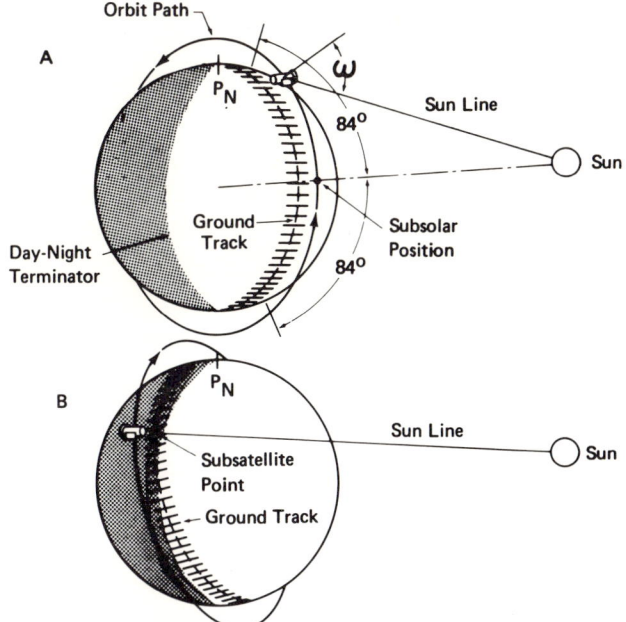

Fig. 1. Meteorological Satellites in Sun-Synchronous Near Polar Orbits

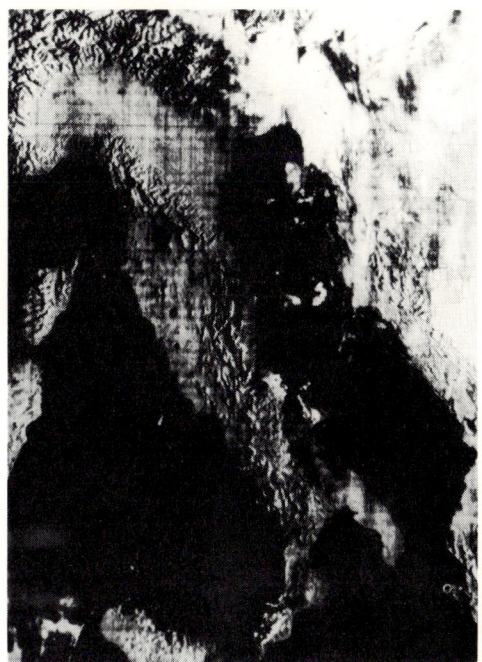

Fig. 2. Early Morning Satellite Image of Italy

Corsica, and Western Italy near Torino where the illumination is at least 5 orders of magnitude dimmer. This picture shows the Apennines and the Alps in pseudo-relief because the early morning sun illuminates only the eastern side of each mountain. Fig. 3 is a 2-mile resolution picture of the Eastern United States taken near midnight three days before the full moon. The clouds and terrain are illuminated by moonlight, while the cities are recorded by their own lights. Thus sensitivity of this nonintegrating single cell scanner is adequate in the 2-mile mode for some night operation.

These pictures illustrate one application well served by a solid-state sensor in a scanner where conventional frame integrating television could not have performed nearly as well. The key is that the sensor is moving with respect to the scene, permitting a different system philosophy than is used in conventional television.

Infrared Scanners

Successful infrared imaging systems have used single detectors in mechanical scanners which perform a full raster scan or in moving vehicle scanners like that just described. Attempts to use more sensitive infrared television camera tubes with full frame integration have been relatively unsatisfactory. The problem is that the thermal photon flux from room temperature objects is large, but the contrast between scene objects is very small, on the order of fractions of a percent for realistic targets.

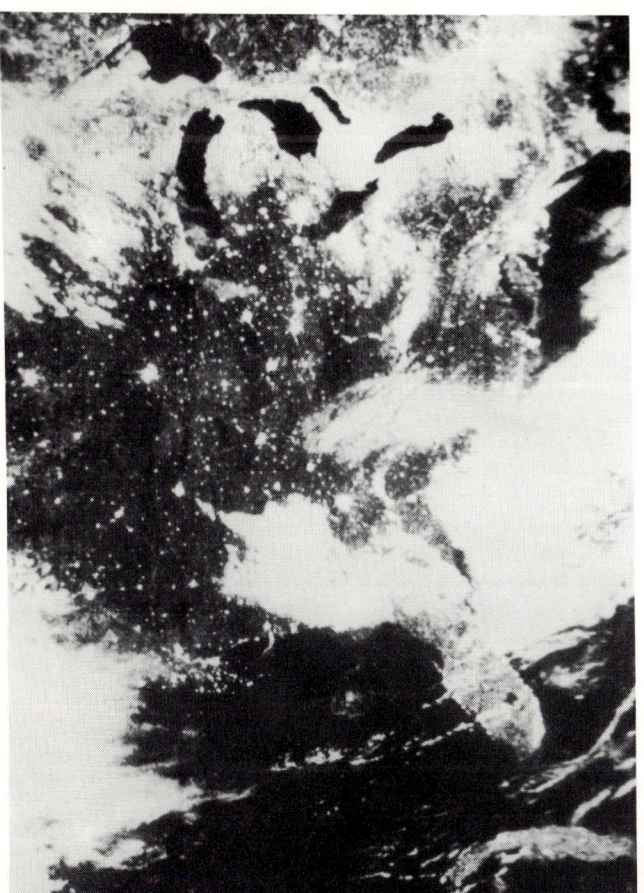

Fig. 3. The Eastern United States at Midnight Three Days Before Full Moon

Thus the infrared imaging sensor does not need sensitivity, but rather the ability to generate useful signals from very small scene contrasts while handling a very large total signal. This requirement implies either extreme uniformity of response across the field of view, or "ac coupling," the ability to ignore the large steady-state average irradiance from the scene while generating a signal from the small point-to-point variations. Normal full frame integrating camera tubes have response or background nonuniformities as high as a prohibitive 10 or 20 percent, and are saturated by the flux from the scene.

In contrast, a solid-state single detector scanning system can have essentially complete uniformity of response over the image area, and use of an ac coupled amplifier ignores the average current from the detector and amplifies only the signal from point-to-point irradiance variations in the image. Of course the detector also reproduces the shot noise associated with the background. To maximize the signal-to-background shot noise ratio one employs a fast optical system to maximize both time varying and background signal, since the shot noise varies as the square root of the latter, cools the lens and detector housing, and uses a detector whose long wavelength response limit falls just beyond the atmospheric spectral window.

To further improve performance, one can use a 100, 200, or more element linear array scanner in which the scene image is scanned perpendicularly to the array. This arrangement, shown in Fig. 4, can be thought of either as reducing the signal bandwidth to band-limit the shot noise, or as increasing the effective sensor area. The signal-to-background shot noise ratio improves as the square root of the number of detectors. Each detector has its own ac coupled preamplifier and senses irradiance variations, ignoring the average background value along its scanning path. However, each line of the image is scanned by a different preamplifier detector combination so channel-to-channel variations in average output signal or in gain can produce line-to-line brightness variations in the reproduced image which may mask the visibility of low contrast detail. Mechanical scanning speed is only one cycle per frame time.

Alternatively, a larger effective sensor area may be provided by placing detectors in series rather than in parallel. As shown in Fig. 4 detectors may be placed one behind the other along a scanning line, using an analog delay line or a CCD to combine the individual output currents in the time delay and integration mode to form a single coherent output. If scanning speed matches the electrical time delay, this scheme gives resolving power equaling a single detector scanner but signal-to-noise ratio is improved as the square root of the number of elements. Further, since each element scans the entire scene, except for the extreme edges, the signal response and background level at the display does not vary from scan line to scan line, but is a function only of scene emittance so uniformity is excellent. The tradeoff here is the need for video bandwidths of 10^7 Hz or higher, and the mechanically difficult design of a high speed x-y raster scanner to provide a flicker-free presentation with large information content.

The common feature of these applications is relative motion of the detector or detector array and the scene image, which both permits and requires an imaging system philosophy unlike that of conventional television. CCD area imagers appear nearly ideal for such applications. An area CCD imager for a moving image is simpler than one for conventional television since it needs no interline transfer register or frame storage register. As shown in Fig. 4, the direction of mechanical motion is horizontal while the higher speed electrical scan lines are vertical. Horizontal electrical scan is synchronized with mechanical scan. Effective exposure time is "dwell time" per element times the number of sensor elements in a row. Each row provides an output once per dwell time, and each output combines the signal from all elements in that row. Therefore, variations in element-to-element response or background current are averaged over the elements in each row, and if the variations are random the variability between row outputs should be much less than for a scanned linear array. Mechanical scanner speed may be as low as one cycle per frame time, and signal-to-noise ratio in a multiplexed output will be higher than that of a single element by N, where N is the total number of elements driving that output terminal. One can vary system gain as a function of mechanical scanning position to accommodate wide variations in image irradiance in the direction of mechanical motion, and sophisticated signal processing schemes can be coveniently implemented by, for example, use of other CCD elements on either the same or separate chips mounted at the focal plane. This type of application appears to be one clearly suited for CCD solid-state imagers, and these approaches should be considered by inventive system designers. Options include not only infrared but also visible image sensing, such as the multi-spectral silicon CCD sensor for NASA's Earth

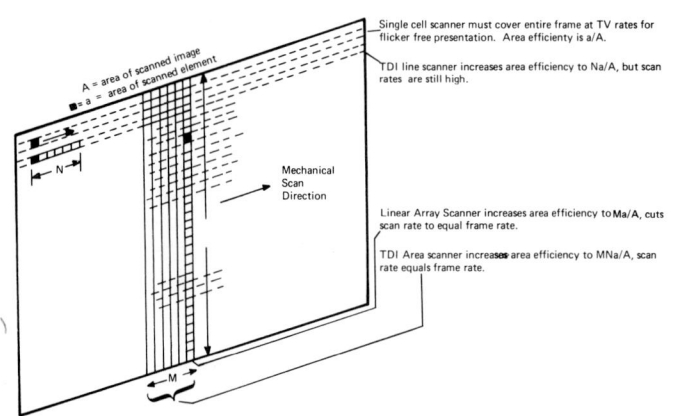

Fig. 4. Scanner May Use Series Elements With Time Delay and Integration to Insure Response, Parallel Elements to Insure Response and Reduce Scanner Rates, or Area TDI for Still Higher Response

Resources Technology Satellite program. Infrared sensor arrays are usually operated with auxiliary CCD's for time delay and integration. Possible new applications include a surveillance camera which is continuously panned through 360 degrees, or a future higher performance weather satellite for observing cloud cover at night.

Photon Counting

Photon counting is particularly important for astronomical imaging, especially in satellite-borne observatories like NASA's proposed large space telescope. For measuring very faint images the astronomer tracks the stars with his telescope to keep the image stationary and uses a long exposure to record enough photon events on each elemental image area to achieve the required radiometric accuracy. Whether measuring star fields, faint nebulosities, or stellar spectra, he usually needs to measure photon fluxes which are only slightly above background, the low contrast problem which seems to occur in most scientific and technical applications. Photon events are random phenomena, and accurate measurement of a small irradiance difference requires recording many independent photon events to differentiate signal from noise, since the measurement accuracy varies approximately as the square root of the number of independent events. For typical low contrast images, one exposes until the background-generated charge is a major fraction of the total charge the sensor can handle without saturating. Hence the ability to measure photon flux above background is limited by the charge storage capacity of the sensor. Accuracy is also limited by sensor noise, usually dark current shot noise and preamplifier noise. For exposures of even a few seconds with a silicon CCD or silicon diode array camera tube the sensor must be cooled significantly to reduce dark current so that neither nonuniformities nor shot noise masks low contrast images.

All of the problems just described occur when the sensor is used in the conventional frame integrating mode. To remove essentially all these limitations the astronomers have pioneered the photon counting mode, using an intensified camera tube or an ICCD as shown in Fig. 5. To ensure that each event can be counted, the photons are absorbed in a photocathode and the emitted photoelectrons are accelerated through 10 to 20 keV before being focused onto the silicon CCD where each electron produces a burst of 2,000 to 4,000 charge carriers. Even though these charge carriers may be distributed over several CCD elements, the resulting output pulse amplitude is many times larger than the sum of dark current shot noise and amplifier noise, and the photoelectron pulses can be discriminated from the noise pulses with a simple threshold circuit which in effect removes the device noise by clipping it. The CCD is then scanned very rapidly so that the whole array is read out 100 or more times per second. If the photon image is so faint that on the average not more than 10 photon events are counted on a CCD element is a second, there will seldom be more than one photon event on the same element in a 0.01-second frame time. Thus the output of the discriminator can be read into a computer or similar device to count the number of events registered on each element over an exposure time of minutes or hours. Very neatly, the limited analog storage capacity of the CCD has been replaced by the almost unlimited counting capacity of the computer, and the combination of the rapid scan rate and thresholding circuit have eliminated problems of dark current in the CCD and device noise. The only dark events which would cause false events are thermionically-emitted electrons from the photocathode, and these can be minimized by modest cooling of the cathode.

As indicated, image sensing by photon counting can also be performed with an intensified camera tube. Calculations show, however, that the noise of an ICCD sensor with DFGA output can be only a few tens to about one hundred electrons, while for intensified silicon diode array target camera tubes the noise is of the order of a thousand or more electrons. Thus a separate added 15-kV intensifier stage would be required with the camera tube, while an ICC will perform photon counting by itself. Further, the fixed geometry of the ICCD permits a direct translation from memory location to image coordinates without need for geometric calibration.

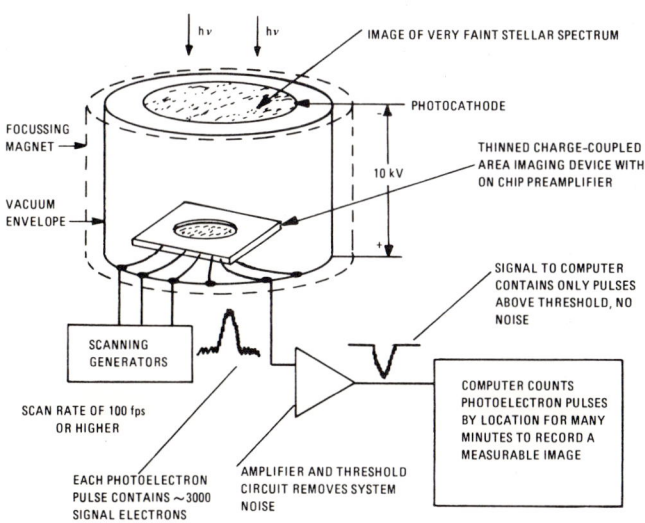

Fig. 5. ICCD is Near Optimum for Photon Counting

Photon counting imaging is therefore another area in which solid-state, CCD, readout excels. Here the advantage is due to the lower readout noise which is enough lower than a tube to avoid a quantum jump in complexity. Applications for photon counting obviously include not only astronomy, but all applications where very faint images must be measured and where a stationary or very slowly varying image permits time exposure techniques. Included would be studies in faint bioluminescence, and spectroscopy of very faint terrestrial sources.

Criteria For Performance Comparisons

Many imaging problems do not provide a clear choice of sensor type like those just described, and either tubes or solid-state devices could be used. To provide criteria for optimum sensor selection, solid-state and electron beam scanned area imaging devices can be compared on the basis of their information handling capability. Use of television techniques to provide quantitative data imposes requirements different from those of entertainment television. The task there is to present a subjectively satisfying flicker free picture within certain bandwidth constraints. For scientific or miliary applications one usually must measure accurately small exposure differences in the presence of large backgrounds and preferably perform these measurements on small details in the image. For accuracy, the sensor must be able to integrate large total signals from each small sampled area so that signal-to-background shot noise and signal-to-sensor noise ratios will be high. For efficiency, each image should contain many independent sample elements and each image should be read out quickly and completely so that new information can be handled. Therefore, the comparison criteria are:

1. Number of independent photoelectron events which can be stored on each sample element.

2. Number of nearly independent sample elements in sensor's field.

3. Data rate in frames per second.

4. Smallness of sensor noise other than background shot noise.

The product 1 x 2 measures the quantity of information which can be handled in each exposure; 1 x 2 x 3 measures the data rate from the sensor; and 4 measures the minimum signal which will give unity signal-to-noise ratio. The fifth criterion, photon-to-photoelectron event conversion efficiency, is considered separately. As will be seen below, these criteria are not entirely independent.

Information Storage Capacity - Number of Sample Elements per Frame

The number of nearly independent sample elements for a solid-state CCD or diode array is clearly the number of CCD cells or diodes. The number of nearly independent sample elements for an electron beam scanned device is not determined by any mosaic structure, but for comparison may be defined using the resolving power as determined by the modulation transfer function. For a diode or CCD array at the Nyquist frequency, one sample diode per test pattern half cycle, MTF = 0.637 as determined by the geometry of an ideal sensor with no gaps between elemental sensitive areas. Reported measured MTF varies from about 50 percent for diode arrays in the visible spectrum, to 0.19 for a Fairchild CCLID-1000B and 0.31 for a T.I. thinned back illuminated 160 x 100 element CCD area array. The CCD values are lower probably because they were measured with tungsten illumination without an infrared blocking filter. Red and near infrared photons are absorbed over a significant distance into the silicon and the carriers diffuse laterally before they are collected. (The CCD results were reported as response to a bar chart, but have been converted to MTF values using Coltman's formula.) Based on the diode results and the expectation of similar results for CCD's, we define a sample element size for a camera tube as corresponding to 50 percent MTF. Fig. 6 shows MTF curves for two Westinghouse EBS camera tubes, computed from published bar

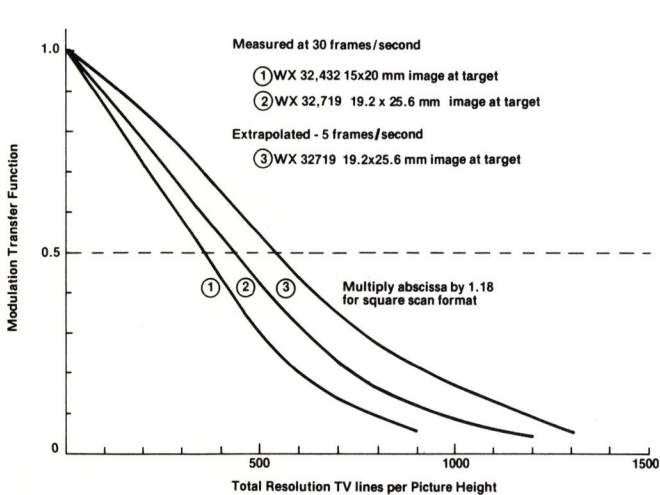

Fig. 6. Modulation Transfer Function for Representative Westinghouse EBS Camera Tubes

chart response data using Coltman's formula. At 50 percent MTF, curve 2 for the WX32,719 at 30 frames per second indicates 430 half cycles in the height and therefore (4/3) 430 = 573 half cycles in the width of a 19.2 x 25.6 mm image at the target. These values are reported in Fig. 7.

Information Storage Capacity - Independent Photoelectron Events per Element

To allow for further electron gun and thermionic cathode improvements, the maximum stored charge for each electron beam scanned sensor is calculated from the maximum signal current, I_{sig}, measured at slower scan rates where the scanning beam does not limit signal amplitude. Then

$$Q_t = I_{sig} t_f / q \quad \text{where} \quad Q_t = \text{stored charge per frame}$$

and t_f = frame time. Q_s, the stored charge per cm^2, is Q_t/scanned area, and Q_e, the stored charge per element is Q_t/number of sample elements. All charge levels are expressed as a number of electrons, since the basic information quantum is the electronic charge. The maximum numbers of electrons per element for CCD sensors were taken from the Fairchild and Texas Instruments reports listed at the end of this paper.

The last line in Fig. 7 shows the CCD charge storage limit of 10^{13} electrons/cm^2 set by the dielectric strength of the silicon dioxide gate insulator, and is due to Barbe. Since the active storage well area for an area array is only 25 to 35 percent of each cell area, for a 25-μm square cell this limit is about 1.9×10^7 electrons per element. The TI back illuminated area CCD approximates 10 percent of this theoretical limit. The Fairchild CCAID244 is lower, because of the smaller area efficiency of the interline transfer array and because of the shallower wells inherent in the buried channel design, although the Fairchild report suggested other less basic causes for these particular arrays.

Sensor Noise Other Then Background Noise

Sensor noise is primarily noise in the first amplifier and associated circuits used to read out the signal. Television camera tubes normally use load circuits and preamplifiers like those illustrated in Fig. 8. In circuit 8b, the signal voltage developed across R_L and C_S feeds a high input impedance voltage amplifier. To provide flat response out to some maximum frequency, Δf_v, the amplifier has a rising gain vs signal frequency characteristic $G = G_o \sqrt{1 + \omega^2 R_L^2 C_S^2}$. Noise sources include the white Johnson noise spectrum of the load resistor, and the preamplifier noise, which is high peaked by the amplifier gain characteristic. The equivalent rms noise current, the integral of the noise spectral density over the video passband, is:

$$I_n = \sqrt{\frac{4kT}{R_L} \Delta f_v + \frac{\varepsilon_n^2}{R_L^2} \Delta f_v + \frac{4}{3} \pi^2 C_S^2 \varepsilon_n^2 \Delta f_v^3}$$

Information Capacity — Elements at 50% MTF

Camera Tubes	Image Size mm	Elements			Q_s el cm^2	Q_t Electrons	Q_e Electrons	E1
EBS WX 32,432	15 x 20	360 x 480		30 fps	1.5×10^{11}	4.5×10^{11}	2.6×10^6	53 μm
WX 32,719	19.2 x 25.6	430 x 573		30 fps	1.5×10^{11}	7.37×10^{11}	3.0×10^6	56 μm
	19.2 x 25.6	530 x 707		5 fps	1.5×10^{11}	7.37×10^{11}	1.97×10^6	45 μm
CCD Fairchild	4.4 x 5.7	190 x 244		60 fps	1.29 to 3.1×10^{10}	3.25 to 7.9×10^9	0.7 to 1.7×10^5	18 x 30
	3.1 x 4.2	100 x 100		30 fps	3.1×10^{10}	4×10^9	4×10^5	31 x 42 (20 x 30 Sens.)
TI	2.29 x 3.66	100 x 160			3.82×10^{11}	3.2×10^{10}	2×10^6	22.5
Theoretical					3×10^{12}		1.9×10^7	25

Fig. 7. Information Capacity Comparison - Elements at 50 Percent MTF for Camera Tubes

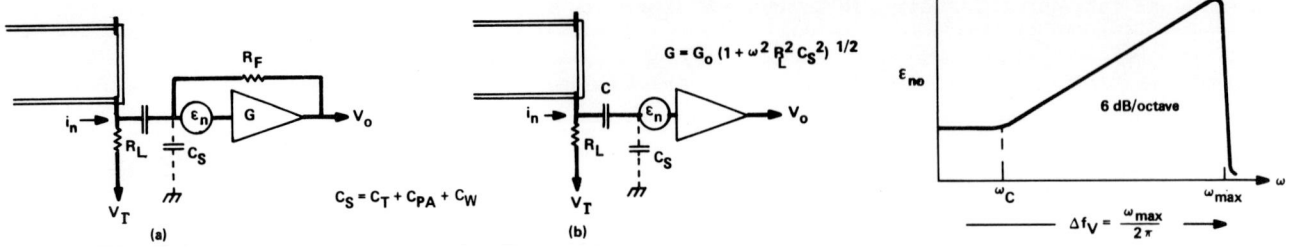

Fig. 8. Television Camera Preamplifier

For a typical camera tube with a preamplifier using two 2N4417 JFET's in parallel, Fig. 9 shows noise as i_n and as electrons per Nyquist sample, q_n. For low system noise, one must minimize the shunt capacitances at sensor output and amplifier input.

Fig. 10 shows the signal circuit for a typical Westinghouse CCD. Each information sample appears as a separate charge packet at the gate of a MOSFET "electrometer" amplifier which drives an off chip operational amplifier. After the voltage change produced by each signal charge is read, the gate node is reset before transferring the next charge packet. Resetting introduces noise. When the MOS transistor reset switch is closed, its Johnson noise appears across C_S, and when it is opened, a sample of this noise, Q_n, is stored on the gate node, adding to the following signal packet. $Q_n = \frac{1}{q}\sqrt{kTC_S}$, and even for a minimal $C_S = 0.17$ pF, $Q_n = 163$ electrons.

To eliminate this noise, Westinghouse engineers devised correlated double sampling (CDS). As shown in Fig. 10, after the gate node is reset, a clamp switch is closed to store on the clamp capacitor a negative voltage corresponding to the noise charge. When the next signal packet is transferred to the gate, the signal voltage at the clamp capacitor output corresponds to signal charge alone, and this signal is sampled for subsequent processing.

CDS also removes much excess 1/f noise since the signal is the difference between two samples taken close together in time. Further, the sample transmitted can be taken when transients have decayed so that switching and clocking noise is eliminated.

With CDS, the major noise contribution, besides leakage shot noise which can be made negligible by cooling, is the amplifier noise which in terms of rms charge per sample is

$$q_n = \frac{1}{q}\sqrt{\epsilon_{n_1}^2 \Delta f_{n1} C_S^2 + \frac{I_{n2}^2 \Delta f_{n2}^2 C_S^2}{g_m^2}}$$

on-chip amplifier noise   off-chip noise

I_{n2} amp Hz$^{-1/2}$ is the equivalent off-chip input noise current, g_m the on-chip MOSFET transconductance, and Δf_{n1} and Δf_{n2} are the effective noise bandwidths. Because CDS samples the noise twice, the effective noise bandwidths are approximately twice the signal bandwidth. To produce a 30 frame per second image with a 100 x 100 element CCD would require 3×10^5 samples/second, and $\Delta f_v \approx 4.5 \times 10^5$ Hz to allow for transients to settle. Δf_n is then about 9×10^5 Hz and for $C_S = 0.25 \times 10^{-12}$F, $\epsilon_{n_1} = 16 \times 10^{-9}$ VHz$^{-1/2}$, $i_{n2} = 2.6 \times 10^{-12}$ AHz$^{-1/2}$, and $g_m = 200$ mho,

Optimum preamplifier design uses large R_L so Johnson noise equals preamp noise.
For $C_T = 10$ pF, $C_W = 2$ pF, paralleled 2N4417 FETs,

$i_n = \sqrt{2(\frac{4}{3})\pi^2 C_S^2 \Delta f_v^3 \epsilon_n^2}$, and $\epsilon_n = 1.2 \times 10^{-9}$ V/Hz^{½}
$R_{eq} = 90 \, \Omega$

Δf_v	10^4 Hz	10^5 Hz	10^6 Hz	4.5×10^6 Hz	10^7 Hz
i_n	1.23×10^{-13}A	3.9×10^{-12}	1.33×10^{-10}	1.18×10^{-9}	3.9×10^{-9}
q_n	38.5 e	122	385	816	1220

Fig. 9. Optimum Preamplifier Design Given $q_n \propto \Delta f_v^{1/2}$, But $i_n \propto \Delta f_v^{3/2}$. Noise is Determined by $C_S \approx 2C_T$

$q_n = \frac{1}{q}\sqrt{1.44 \times 10^{-35} + 6.08 \times 10^{-36}}$
= 28 electrons.

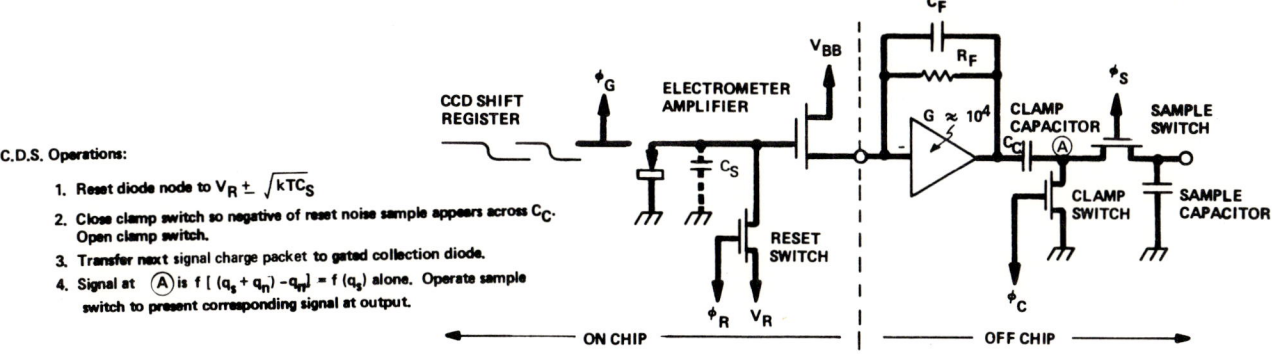

Fig. 10. Correlated Double Sampling Removes Reset Noise at CCD Output

Thus at modest data rates, the system noise of a CCD with correlated double sampling is low. To scan a full television format at 30 frames per second would require at least a 7-MHz data rate for the Fairchild CCAID488 which provides interlace by scanning the entire image 60 times per second and discarding one field in every frame. With correlated double sampling the noise bandwidth would then be 14 MHz and q_n = 112 electrons, or I_n = 0.25 x 10^{-9}A. At this data rate, Fairchild uses a distributed floating gate amplifier, and reports Q_n = 10 to 20 electrons even at 7 MHz.

As with tubes, shunt capacitance must be minimized for low amplifier noise. But with CCD solid-state arrays, C_S can be two orders of magnitude smaller than with tubes. Because of the variation of e_n with g_m and therefore with C_S, the system noise varies roughly as $\sqrt{C_S}$, and is about an order of magnitude smaller for the CCD at equal scan rates.

Other Criteria - Quantum Efficiency and Data Rate

Fig. 11 shows conversion quantum efficiency, electrons per incident photon, as a function of wavelength for several sensor types. For a wavelength of 555 nm in the center of the visible spectrum, an antireflection coated photodiode array can reach about 94 percent, a very good back illuminated thinned CCD or a silicon diode array target camera tube about 82 percent, a typical thinned CCD 40 percent and a specific Fairchild CCD with illumination through semitransparent gates and an interline transfer structure which reduces area efficiency to 40 percent, about 15 percent. The yield of a very good S-25 photoemissive cathode is about 10 percent. To determine the signal of a real application one must integrate the product of the photon flux with the quantum efficiency at each wavelength over the spectrum, but the comparison at that single wavelength shows that a good silicon photodiode sensor can provide nine times the signal of a good photocathode, and silicon CCD sensors up to eight times the signal. These curves illustrate why work is being continued to further reduce noise in CCD arrays to take advantage of this high quantum efficiency. In fairness, it must be noted that the data on the TI arrays is taken from report published in March 1975, and the Fairchild curve was extrapolated from another similar Fairchild array at about the same date and both may now be too pessimistic.

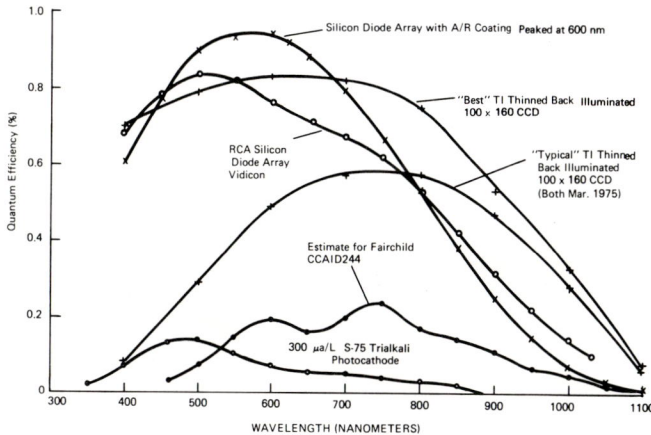

Fig. 11. Quantum Efficiency vs Wavelength for Relevant Solid-State and Electron Beam Scanned Sensors

These curves also show why the TD1-CCD imaging sensor is important. Since there is no need for an interline transfer structure, area efficiency can be nearly 100 percent and quantum efficiency at least like that of the "typical" TI array should be readily achievable.

The data rate in independent frames per second is not limited by lag effects in any solid-state array sensor as it is in a camera tube. Lag effects in CCD's produce transfer inefficiency and may limit resolution, but each frame is read completely. Instead, the data rate limitation is usually the amplifier noise, which was discussed in the preceding section. Thus the data rate criterion is not independent, but is absorbed into criterion 4, and will not be treated separately.

Conclusions

The table of Fig. 12 lists relative advantages of solid-state sensors and of electron beam scanned sensors at the state of the art in late 1975 or early 1976, using numbers generated in the previous sections. Camera tubes provide a larger information content per frame, principally because the silicon arrays are larger in camera tubes, and the small scale picture informity is better because the "resolution element" is defined by the cross section of the electron beam which typically covers many diode elements. CCD imagers are especially attractive for the large ratio of maximum signal per element to system noise, for complete freedom from lag at all exposure levels, for low system noise which permits low light level operation intermediate between a silicon vidicon and an EBS/SIT tube, and for metricity, the fixed image geometry that is so useful in quantitative measurements. As new CCD imagers are developed the comparison results may change, and the comparison methods suggested here should be useful in updating the comparison.

Camera Tubes	CCD Imagers
More total resolvable elements/frame	Higher SNR at maximum signal
Better small scale response uniformity	Large effective dynamic range (except for blemishes)
Large single terminal output data rate	Lower system noise
Lower blooming	Freedom from lag at all light levels
Larger Image Size	Geometric Fidelity
High device image quality	Metricity
	Compactness
	Long stable life
	Low voltages
	TDI mode available

Fig. 12. Comparison Table

All the foregoing comparisons, however, ignore device quality factors. Camera tubes represent a relatively mature art, and picture quality is high. The visibility of televised images from CCD sensors, however, is set by factors other than the total number of elements or amplifier noise. Leakage current nonuniformities, "fat zero" noise and nonuniformities in surface channel devices, and element to element response nonuniformities can all produce either an interfering pepper and salt pattern or a pattern of striations as light and or contrast is reduced, and today the extremely large dynamic ranges predicted for these devices are seldom realized experimentally. Because the eye integrates in space as well as in time, bar patterns images are visible with signals of only 30 or 40 electrons per element, but the entertainment quality images reported in the literature are usually made with signals not less than 10 percent of the maximum saturated value. Thus, it seems unlikely that solid-state imagers will soon replace camera tubes in commercial television. But there is every reason to consider them seriously for either compact cameras of modest performance, or for applications like those described at the beginning of this paper where scene motion provides a real advantage rather than a handicap.

References

Antcliffe et al, "Large Area CCD Imagers for Spacecraft Applications," in Proceedings of Symposium on Charge-Coupled Device Technology for Scientific Imaging Applications, Mar. 6-7, 1975, Jet Propulsion Laboratory, Pasadena.

Barbe, D. F., and Schmidt, W. A., "Infrared Charge-Coupled Imagers," IRIS Imaging Speciality Group, 28-29 November, 1972.

CCD Photosensor Array Development Program (Phase II), Final Report, April 1975 on contract N00039-73-0015 to Naval Electronic Systems Command by Fairchild Space and Defense Systems. The work reported was completed by 30 August 1974.

Coltman, J. W., "The Specification of Imaging Properties by Response to Sine-Wave Input," JOSA44(6), 468-471 (June 1954).

PYROELECTRIC VIDICON THERMAL IMAGER

E. H. Stupp

Philips Laboratories, Briarcliff Manor, N.Y. 10510

Abstract

This paper will review the operating principles, state-of-the-art performance, applications and future improvements for pyroelectric vidicon thermal imaging systems.

Infrared imagers have long held the promise of wide variety of military, medical, commercial, industrial and civil applications. These systems convert the self-radiated IR emissions that are characteristic of all objects to visible wavelengths for viewing in real time or from a permanent record. Many of these IR imagers are based on arrays of quantum detectors, operating at cryogenic temperatures, and are sufficiently expensive to be uneconomical for many applications.

Within the last few years, a new type of IR imaging device — the pyroelectric vidicon (PEV) — has been developed. This tube operates at room temperature in a modified video camera. PEV systems are already available in small quantities at a fraction of the cost of quantum detector systems, with still lower prices being projected for production quantities.

In this paper, the principals of operation of the PEV will be reviewed. The state-of-the-art performance of PEV systems will be presented, together with some of the applications being investigated. Finally, the projected performances expected with new target materials and advanced designs will be summarized.

Pyroelectricity is a phenomenom associated with non-centrosymmetric crystals exhibiting a spontaneous polarization. A change of the temperature of the crystal results in a change in the surface potential of the crystal. In a vidicon (Fig. 1) the electron beam senses the change of potential and deposits a charge on the crystal surface to reestablish an equipotential at approximately cathode potential. This deposition of charge is the video signal.

DIFFERENCE BETWEEN PYROELECTRIC VIDICON AND CONVENTIONAL VIDICON

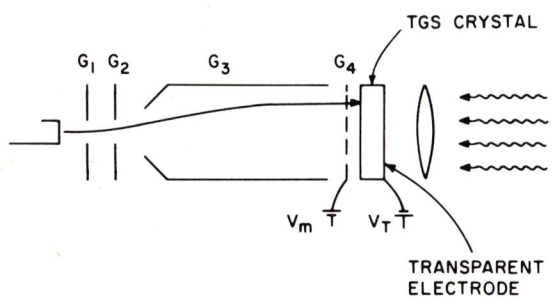

Figure 1

A. PYROELECTRIC VIDICON CAN ONLY RESPOND TO TIME VARYING SIGNALS

B. PYROELECTRIC MATERIALS ARE EXCELLENT INSULATORS NOT CAPABLE OF LEAKING THE CHARGE DEPOSITED BY THE BEAM

CONSEQUENCES

A. TO OBSERVE STATIC SCENES ONE MUST CHOP THE INCIDENT RADIATION OR PAN THE SCENE

B. IN ORDER TO MAINTAIN OPERATION IN C.P.S. ONE MUST PROVIDE A PEDESTAL CURRENT

If the window of the tube is an infrared transmissive material, the change in temperature can be produced by an IR image focussed onto the target. Since the tube responds to radiant power, a signal is produced for all infrared wavelengths absorbed. The most common PEV target material is triglycine sulfate (TGS). With this material, PEV tubes with appropriate windows will image from 2 μm to 400 μm. All the data presented in this paper was taken in the 8-14 μm spectral range.

Once the surface potential has been reestablished at cathode potential, no more electrons can land despite temperature variations in the target produced by the image of the scene. To continue to observe a signal, a change in temperature must be produced which results in a change in surface potential. Two techniques which are employed to achieve this time varying temperature are: (1) chopping the incoming radiation and (2) panning the image of the scene across the target.

Pyroelectric materials are extremely good insulators. An electron beam landing at low energy, as in the vidicon, is an almost perfect rectifier. The combination of these properties results in a condition in which there is a net buildup of charge with time. Eventually, no further charge will be able to land if a mechanism is not provided for raising the surface potential by removing this excess charge.

Several techniques have been developed for providing this potential increase. At Philips Laboratories, an all electronic technique is employed. In this approach (Fig. 2), the target potential is made higher than the first crossover voltage during beam flyback by pulsing the cathode negative and the beam is turned on. Under these conditions, the secondary emission coefficient is greater than unity during flyback and thus more electrons leave than land during this time, resulting in an increase in surface potential or positive "pedestal".

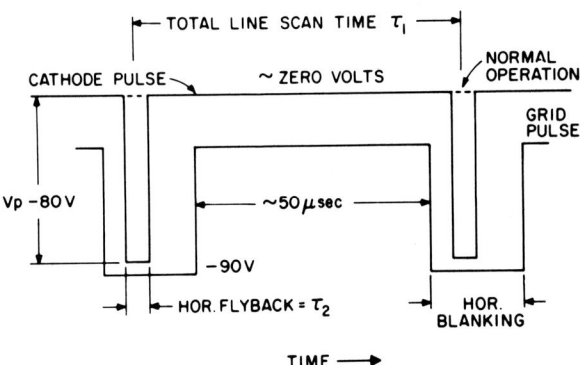

Figure 2

In standard vidicons, an electron gun is used in which there is a crossover in the triode. This results in considerable "heating" of the electron beam. Use of such a hot gun in the PEV results in increased minimum resolvable temperature (MRT). Figure 3 shows the MRT of a TGS PEV with such a conventional gun compared with MRT of a new PEV with a non-crossover gun. This latter type is characterized by a much lower beam temperature (LBT) even at high pedestal currents. At 100 nA pedestal, the LBT PEV has a pan mode MRT of less than 1°C at 250 TVL (6.25 Lp/mm). At 200 nA pedestal, the MRT reduces to 0.6°C at 250 TVL. At low spatial frequencies, the MRT is significantly below 0.1°C.

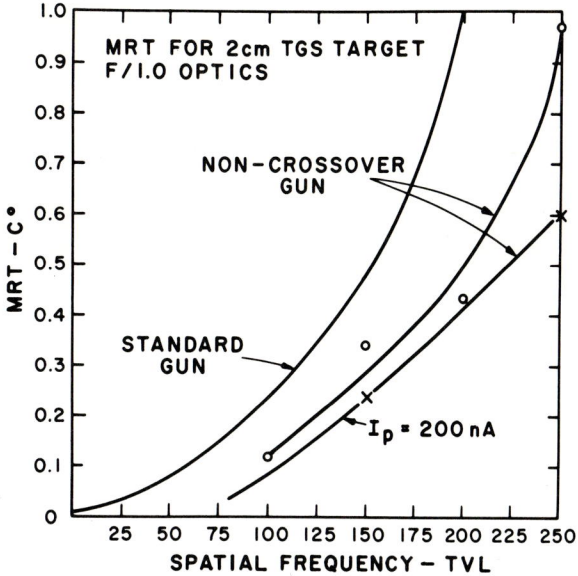

Figure 3

With current fabrication technology, long life is routinely being achieved (Fig. 4). No change in gas pressure in the tube or MRT (2.5 Lp/mm) is observed after 3000 operational hours. At this time, no specific failure mode has been identified.

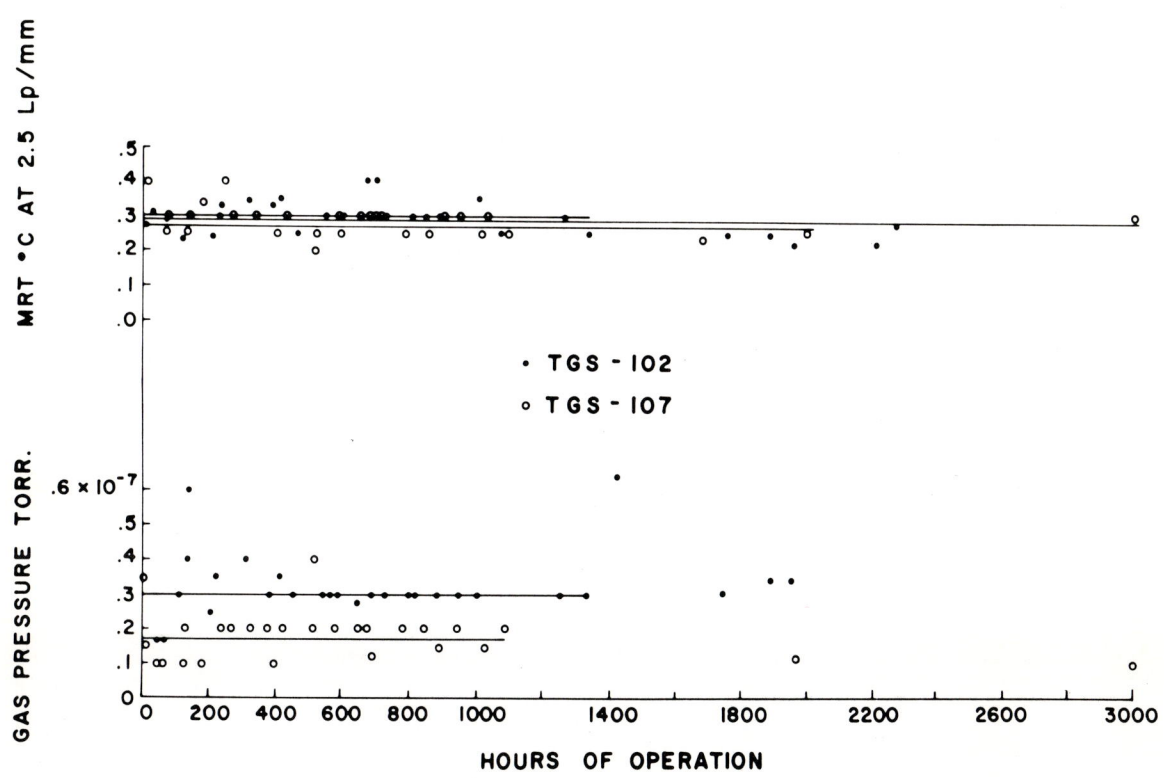

Figure 4

An important application of PEV systems being investigated is area security. Since the PEV system does not respond to static scenes, it can be used both as an alarm sensor and an imaging sensor. In the absence of any motion in the scene, no signal is generated. If any object moves in the scene, an alarm can be generated without human intervention. Recognition of the object is obtained by switching to an imaging mode - panned or chopped. Humanoid recognition is obtained at a range of 200 m with a 100 mm FL lens.

Another application for the PEV is fire location. While it might seem that a source as hot as a fire could not be difficult to find, smoke obscuration, particularly in enclosed areas, often makes location of the burning source problematic. The increased smoke penetration available at long wavelengths, together with the high sensitivity of the tube, have provided very encouraging results for this application. In a similar use, the PEV has been employed in fire location in a forest environment.

Medical diagnostics is another application of the PEV receiving attention. For this use, it is desirable to operate in the chopped mode and, with appropriate signal processing, provide a permanent or semi-permanent display. One processing scheme being investigated utilizes a storage tube for integrating up to 100 frames. The MRT improves as $\sqrt{N}$, where N is the number of frames integrated (Fig. 5). The venal system in the arm can be easily resolved with 1/2 second exposure with this system, indicating better than 0.1°C resolution.

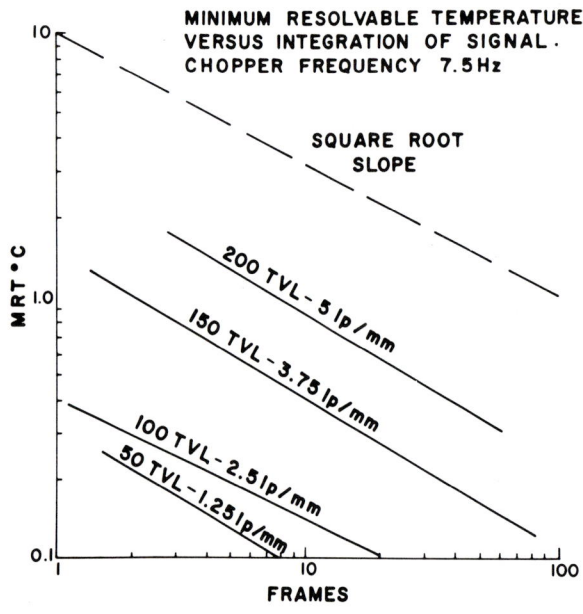

Figure 5

Future pyroelectric systems may employ tubes with even better resolution and MRTs. Deuterated triglycine fluoroberyllate (DTGFB) is a target material currently being studied for PEV application. Figure 6 shows the measured MRTs at 100 nA pedestal currents of a DTGFB PEV compared with a TGS PEV. Significant MRT improvements are observed for all spatial frequencies. The most dramatic decrease in MRT occurs at the higher frequencies.

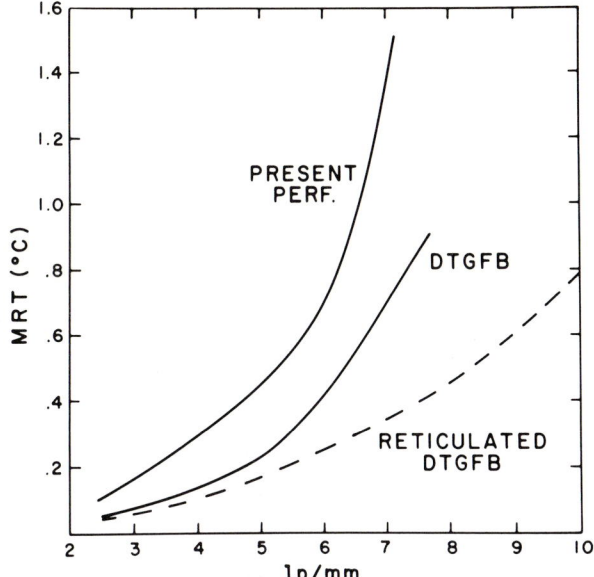

Figure 6

Also being investigated are techniques for reticulating these targets. This procedure reduces the thermal diffusivity of the target, with a resultant improvement in resolution. Theoretical calculations indicate that reticulated DTGFB PEV systems will perform competively with the medium performance FLIR.

Acknowledgment

The work at PL was partially supported by USAMERDC, Fort Belvoir, Va., Contract No. DAAG53-75C-0256 and DAAG 53-76C-0053.

Bibliography

1. T. Conklin, B. Singer, M. H. Crowell, R. Kurczewski, IEEE, IEDM, Washington, D.C., 1974.

2. T. Conklin, B. Singer, IEEE, IEDM, Washington, D.C., 1975.

3. L. E. Garn, F. C. Petito, IEEE, IEDM, Washington, D.C., 1973.

4. S. C. Hayden, J. F. Balascio, G. M. Loiacono, Proceedings EOSD Conference 1974 West, San Francisco, Ca., 1974.

5. C. N. Helmick, W. H. Woodworth, IEEE, SAF, Albuquerque, N. M., 1975.

6. Kourtz, Pinson, Report FF-X-52, Forest Fire Research Inst. Canadian Forestry Inst.

7. R. Kurczewski, R. Levitt, SPIE, San Diego, Ca., 1975.

8. F. C. Petito, L. E. Garn, et. al., IEEE, IEDM, Washington, D.C., 1974.

9. B. Singer, M. H. Crowell, T. Conklin, IEEE, IEDM, Washington, D. C., 1973.

10. B. Singer, M. H. Crowell, T. Conklin, Device Research Conf., Santa Barbara, Ca., 1974.

11. B. Singer, J. Lalak, M. Jennings, IEEE, SAF, Albuquerque, New Mexico, 1975.

12. B. Singer, Electro-Opt. Systems Design, p. 30, July 1975.

13. E. H. Stupp, M. H. Crowell, B. Singer, IEEE, IEDM, Washington, D.C., 1972.

14. R. Watton, IEEE, SAF, Albuquerque, N. M., 1975.

GENERAL APPLICATION OF MICROCHANNEL IMAGE INVERTERS

Jon Tegethoff and Ferd Fender
Ni-Tec, Inc.
Skokie, Illinois

Abstract

The basic purpose of this paper is to review the performance improvements that have been made over the past few years and to highlight the strengths and weaknesses of the present microchannel plate (MCP) inverter. On reviewing these strengths and weaknesses, present applications will then be surveyed and a few potential applications will be highlighted.

Introduction

Most of the strengths and weaknesses of microchannel image intensifiers lie in the primary gain element of the intensifier, the microchannel plate (MCP). The MCP is a thin glass structure with many holes or "channels" spaced approximately 15 microns apart. One electron entering the input end of the MCP can produce as many as 10,000 or 20,000 electrons at the output surface. The MCP electron gain is easily controlled by varying the voltage across the MCP. Varying the MCP voltage therefore varies the overall intensifier gain. One of the key features of this type of intensifier is that the resolution (MTF) is not affected when the gain is changed in this manner.

Present Performance

When the second generation microchannel intensifier was first developed, the existing operational performance of the three stage first generation devices were used as the standard for operational performance, especially in the areas of luminous gain and modulation transfer function (MTF). An ion trap is presently employed in all MCP (inverter type) intensifiers and consequently luminous gain has not been a problem. In the area of MTF considerable improvement has occurred over the years, as can be noted in Fig. 1.

	(cycles/mm)					
	2.5	5.0	7.5	10.0	12.5	15.0
3 stage first generation spec.	90%	77%	60%	44%	32%	23%
1971 typical	83%	64%	51%	40%	25%	20%
1974 typical	92%	80%	62%	45%	36%	32%
Present typical	93%	80%	66%	52%	40%	32%

Fig. 1. Modulation Transfer Function

The increase in the last year is due to a decrease in the average center to center spacing of the MCP from 15.5 microns to 14.5 microns. These measurements are for standard production 25 millimeter image intensifiers. By further decreasing the center to center spacing to 10 microns and making improvements in the proximity focusing section high resolution versions of both 25 millimeters and 40 millimeters have been produced (Fig. 2, high resolution and present 25mm). The improvements in the proximity section result from a tradeoff for gain, with the high resolution tubes being capable of maximum luminous gains of only 5000. These higher resolution versions are non-standard and therefore higher priced.

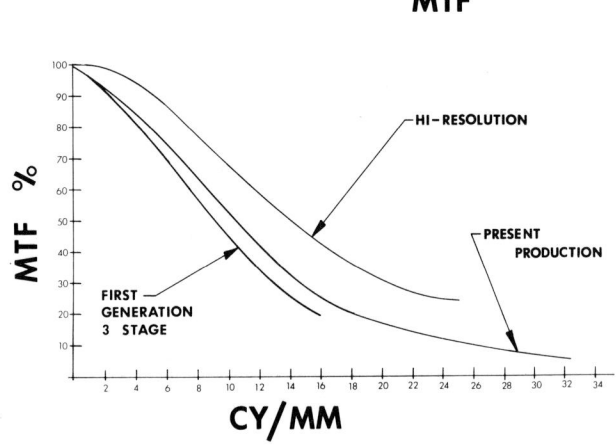

Fig. 2

Although dramatic changes have occurred over the past ten years with respect to photocathode sensitivity, no significant changes have occurred in the last three years. (Fig. 3.).

One of the greatest advantages second generation devices have over first generation is the containment of bright point sources. This property of containment has led to a misunderstanding by some users as can be shown by quoting a tentative specification from an informed user.

"Point Automatic Brightness Control: When the full useful area of the photocathode is uniformly illuminated at 10^{-6} foot candles and a 0.3mm disc illuminated at 95 ±5 foot candles is moved anywhere in the useful area, the maximum output brightness (Luminance) anywhere on the phosphor screen must not exceed 10 foot lamberts."

This requirement is contrary to present MCP characteristics. Since the tube without a microchannel plate has a gain of about 10, the MCP would have to have an attention of one hundred to meet this requirement. This is an order of magnitude different than obtainable from a standard MCP intensifier with the electron gain of a channel varying from 5000 in an unsaturated mode to .1 in full saturation. Beyond full saturation (50-100 f.c.) the electron gain again increases. This region is not a problem since in normal circumstances enough current is drawn by the phosphor screen that the automatic gain control circuit of the power supply becomes operative.

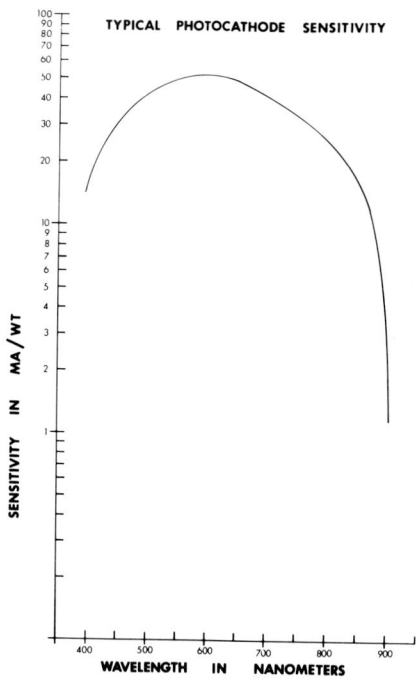

Fig. 3.

Containment is really a two part problem. Intensity attenuation has been previously discussed. Size containment is inherently good with a microchannel plate due to the lact of interaction between adjacent channels, see Fig. 4. These containment characteristics are good enough to photograph an image where the scene brightness is 10^{-5} F.C. and contains point sources of 50 to 100 F.C., a dynamic range of about 10^7.

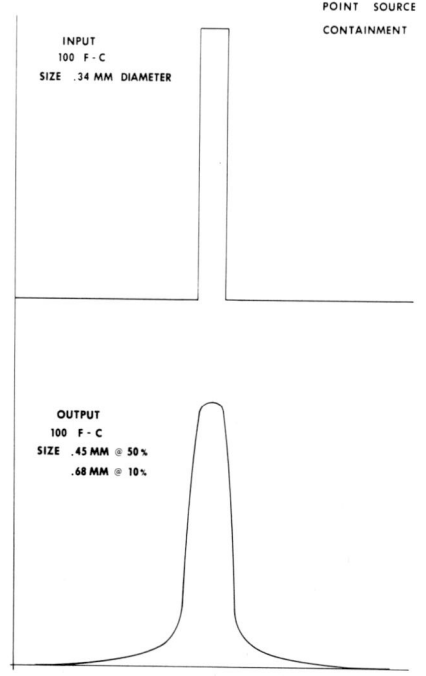

Fig. 4.

Applications

In the future as has been true in the past, the greatest majority of MCP intensifiers will be used in military applications. The large numbers required by the military will greatly reduce the price and open up new applications. The military price has been reduced by approximately one third in the last year alone. Since commercial prices have also been reduced by approximately twenty-five percent in the last year a number of new applications which previously used first generation devices are now using second generation devices.

Airport surveillance of baggage has become extremely important over the last few years. There are a number of different approaches used in X-ray examination of luggage which have been successful. Fig. 5 is one of these successful approaches.

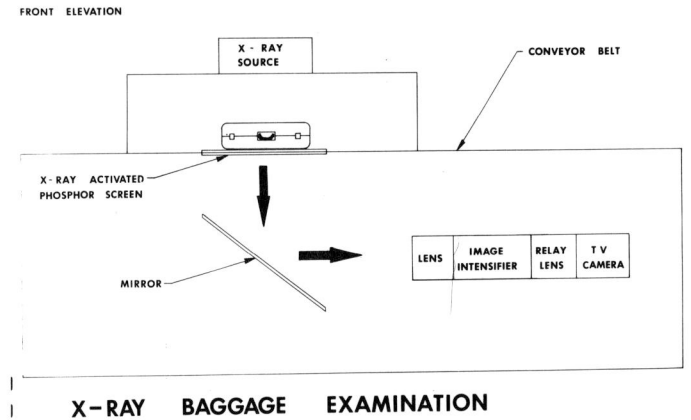

Fig. 5.

The initial system design used a first generation image intensifier and had a system resolution which limited observations to 24 gauge wire. The redesigned system using a second generation intensifier is capable of discerning 28 gauge wire which is half the size of 24 gauge wire.

Nighttime photographic applications are improved by using the second generation intensifiers since the peripherial distortion is greatly reduced and improves the "effective" photographic field of field - the nighttime photographers do not have to center the equipment on the action. This improves the probability of capturing the action on film or video tape, especially when a group of people are under surveillance.

Early last year the Department of Agriculture purchased standard direct view units (See Fig. 6). They have been used with partial success by an entomologist of the Department of Agriculture in the study of moths. A number of types of moths are being studied to understand the mating characteristics. Of particular interest is a moth (Pectinophora gossypiella) which lays eggs on cotton, the larva of which causes extensive damage. The intent of the study is to learn enough about these nocturnal mating habits to be able to interfere. Lights of course do interfere and the study is impossible without the low light level devices.

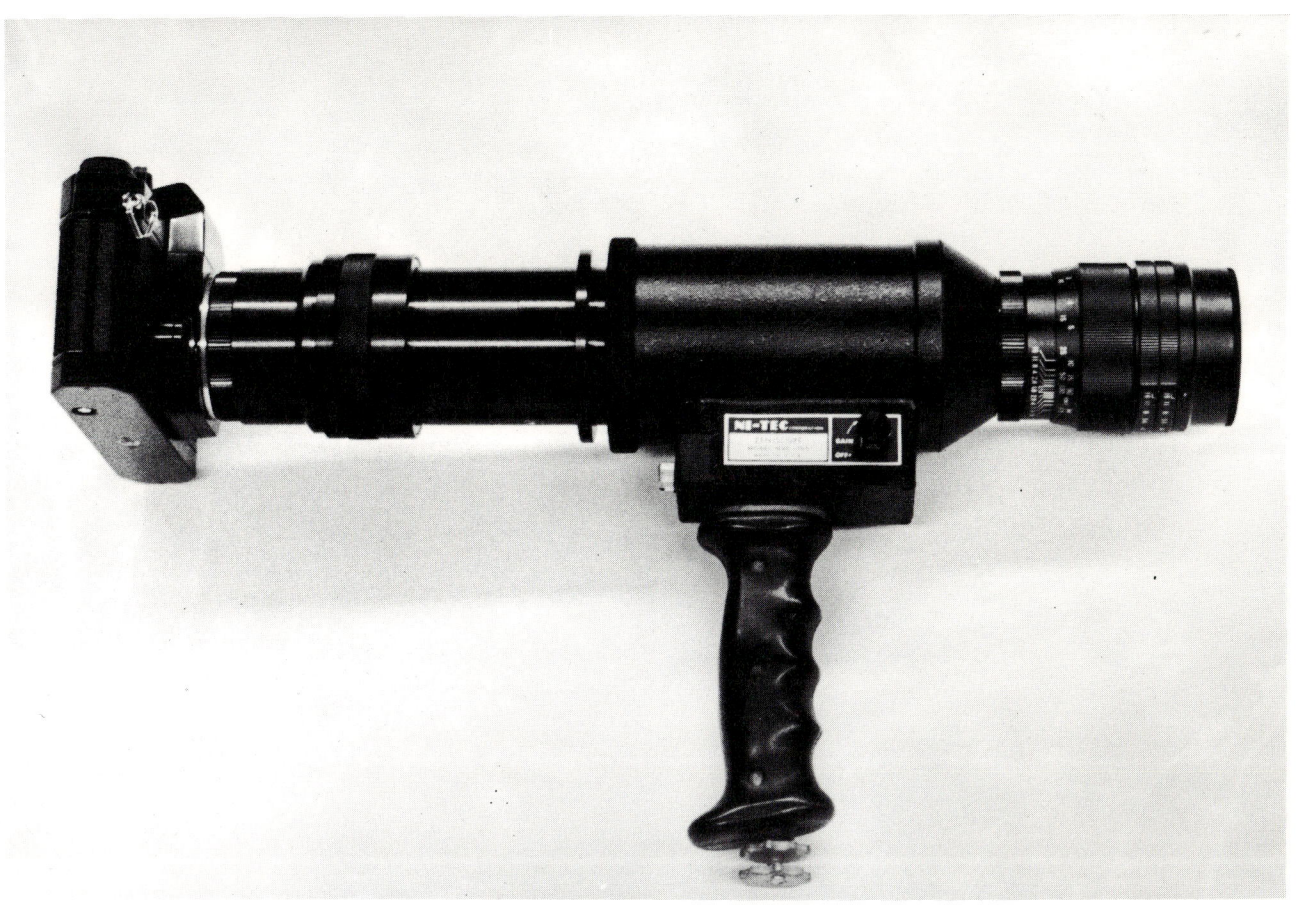

Standard Direct View Unit with 35mm Camera

Fig. 6

Standard MCP image intensifiers are readily adapted to gated applications. In this mode of operation, the tube is generally turned on for only brief periods of time to limit or control the time that optical images are transmitted through, and amplified by, the intensifiers.

Speed of turn on and turn off transition is generally limited only by external circuits controlling the gating voltages. The tube itself is essentially limited only by the electron transit time within the tube. (5 nanosecond rise and fall times have been achieved in the laboratory using delay-line discharge techniques.)

Practical gated power supplies have been delivered for use with pulsed laser systems. These have rise and fall times somewhat less than 100 nanoseconds. Repetition rates are up to 16 KHZ. These pulsed laser-intensifier systems act like an "optical radar". A rough (± few hundred feet) estimate of range to a target may be determined, but more importantly, backscatter from intervening atmospheric haze is effectively suppressed. Detection, identification and tracking of targets has been achieved at ranges of several miles.

Commercial applications of gated laser systems are restricted by governmental safety laws which greatly limit laser power. These restrictions would require some relaxation before a commercial marketplace can be realized.

Similar gated systems have been utilized with a narrow band-pass optical filter for daylight tracking of laser illuminated targets. The fact that the intensifier is off most of the time greatly reduces the normal daylight background.

Slower gating rates or single pulse systems are useful for photographic applications. Here the output image is directed to a shutterless film transport by a relay lens or, more effectively, by direct film contact with the output fiber optic surface. 25mm systems with pulse widths or "on" times from 5 milliseconds to continuous "on" and repetition rates from 0 to 100 pulse/second have been delivered for this purpose. 40mm prototypes have been constructed. It should be noted that the systems delivered do not gate off the screen or MCP voltage. This means that the film may be fogged by slight intensifier "noise" if left in place for any great length of time. Alternatively if about a 1-second "ready" signal is available, the screen and MCP voltage may be applied for only a short time prior to actual gating on and image intensification. In repetitive photographic applications wherein the film is in place for only short periods of time, this effect is of little consequence.

The predominate present commercial use for both direct view and TV and photographic intensifiers is by police and law enforcement agencies for surveillance, nighttime recording, and weapon firing in attempting to control narcotic traffic, terrorism and other criminal activities. Second generation has been successfully used for general surveillence and recording video tapes or photographs for prosecuting cases in court.

Session 2
APPLICATIONS OF LOW LIGHT LEVEL DEVICES IN PUBLIC SERVICE

Session Chairman
William Hawley
Night Vision Laboratory

DETECTION OF MINE HAZARDS WITH INFRARED IMAGERS

Raymond M. Stateham
U.S. Bureau of Mines, Denver Mining Research Center
Building 20, Denver Federal Center
Denver, Colorado 80225

Abstract

Infrared imagers have been used to remotely sense several potentially hazardous conditions in mining operations; among these are loose rock, misfires, shorted power cables, overheated bearings, and combustion (both existing and incipient). In addition, the potential use of infrared imagers as mine rescue tools has been evaluated.

Introduction

Any instruments or techniques that are capable of detecting hazardous conditions from remote, safe locations can be used to improve safety in the environment in which they are used. Infrared imagers meet this requirement and have been evaluated for the detection of several unsafe or undesirable conditions associated with mining operations. These undesirable conditions range from loose rock to fires beneath the earth's surface, and include such hazards as faulty electrical equipment, hot bearings, and misfires; their kinship, for purposes of this report, is in the fact that they can be detected by measurement of anomalous temperatures, and that the same or similar instruments have been used for their detection.

Basic concepts that are inherent to these studies are not new. For example, the fact that the surface of loose rock differs in temperature from the surface of adjacent solid rock has been known for several years.[2,7] Furthermore, it has been known that the earth's surface over buried fires was warmer than the surrounding terrain because of the melting snows. However, it was only in the last few years that instruments capable of detecting the small temperature differences involved became available. In fact, the infrared imagers used in these investigations either were developed or were first made available for civilian use by the military after these investigations began. With these portable, sensitive instruments, it has been possible to identify several types of hazardous conditions because of associated thermal anomalies. The results of several investigations, conducted with infrared imagers, are summarized in this report.

Acknowledgements

The Bureau of Mines appreciates and acknowledges the assistance of the Department of the Army for the development and declassification of the Army's infrared imager used in this study. Acquisition of the data necessary for this report would have been impossible without their cooperation.

Instrumentation

Three basically different infrared imagers have been used in various phases of this investigation: Philips Audio Systems' Handheld Infrared Scanner*, AGA Corporation's Thermovision, and Hughes Aircraft Corporation's Probeye. Most of the work was done with the Philips unit or its military predecessor because this unit was first available and was first approved for use in coal mines.

All of these instruments are lightweight, small in size, battery powered, and have a sensitivity of about 0.2^0C in a spectral region of 3 to 5 micrometers (μm). They have been described in detail in other publications.[1,5-6]

In addition to the imagers, several more conventional instruments were used to measure "ground truth" information such as air and rock temperatures or temperature differences. These instruments included radiometers, mercury thermometers, and probes tipped with thermocouples or thermistors.

---

*Reference to specific makes or models of equipment is made to facilitate understanding and does not imply endorsement by the Bureau of Mines.

Investigations

Loose Rock Detection

Loose rock detection was the subject of several published articles [1-4,7] and is briefly reviewed here for completeness. The concept of detecting loose, hazardous rock by means of temperature differences between loose and solid rock is based on the premise that such a temperature difference will exist, provided a contrast exists between the temperature of the air in the mine and the temperature of the rock on the surface of the underground opening. The fracture or parting behind a piece of loose rock should insulate the loose rock from the rock mass, thereby attenuating the heat transfer between the loose material and the rock mass. Consequently, the piece of loose rock should be more responsive than the solid rock, to the temperature of the mine air, and a small temperature difference should exist between the loose and solid material.

The initial tests were designed to establish the existence or nonexistence of such temperature differences and, if possible, to determine the rate of development. Tests were performed in mines where the ventilation could be controlled to meet the needs of the experiment. All ventilation was stopped in these mines for one or more days so that the temperature of the surface of the opening and the temperature of the air reached equilibrium. To verify that such equilibrium existed, measurements were made through the mines (while the mine air was not being circulated), and no temperature differences were detected between loose and solid materials. The ventilation systems were then placed in operation, and comparisons between known loose rock and adjacent solid rock were made at specific locations at regular intervals. The measurements indicated that temperature differences developed between loose and solid rock within 15 minutes. Although no temperature differences existed before the ventilation systems were placed in operation, temperature differences from 1^0 to 7^0C developed in rock near the intake air. At greater distances into the mine, the temperature differences between loose and solid rock were 0.2^0 to 1.0^0C and were dependent on the temperature of the outside air, the volume of air, and the distance into the mine.

In large, extensive mining operations, the mine air and the surface of the rock at large distances from the intake air are probably near equilibrium and temperature differences between loose and solid rock may be nonexistent or too small to be detected with present instrumentation. However, loose rock can be detected at the working faces because (a) although the air temperature and the rock surface temperature are in equilibrium, a temperature difference usually exists between the air and the rock at depth[1], and (b) the gradient of such a difference should be nonlinear with the change greatest near the surface of the opening. Consequently, at the working face, a new surface is exposed and temperature differences develop between loose and solid rock in the new surface. Because most rockfall-related accidents occur near the working face, the size of the mine is not a limiting factor in the usefulness of infrared techniques for loose-rock detection.

Because the initial field studies indicated a feasibility of detecting loose rock by infrared techniques, followup studies were made in several mines with different rock types and mine environments. The success of these investigations varied with (a) the magnitude of the air-to-rock temperature contrast, and (b) the severity of the mine environment, such as temperature and humidity.

Temperature differences between loose and solid rock have been observed with the scanner or measured with the infrared thermometer in a number of mines in several types of igneous or sedimentary materials, including most rock types and coal. With the exception of coal, which has a lower conductivity, the material or rock type apparently is not a major influence on the magnitude of the temperature differential. Rather, the temperature differential appears to be primarily influenced by the magnitude of the air-to-rock contrast and by the amount of air moving past the rock surface.

In most mines, the normal loose-to-solid rock temperature difference is from 0.2^0 to 0.5^0C. The Burgin mine near Eureka, Utah, was an exception to this general rule in that the usual temperature difference in this mine was from 1^0 to 3^0C. This larger temperature difference was caused by large volumes of fresh air required to cool work areas, which were heated by hot water (about 80^0C) flowing from the rock. Data collected from the mine studies during the investigation are summarized in Table 1. Data shown in Table 1 were collected in 60 different underground openings; 26 of these were coal mines. The category "others" include mines producing silver, tungsten, and uranium in addition to mines used for research, a highway tunnel during excavation, and an underground facility used for explosive forming.

Table 1. Summary of Loose Rock Detection Data

Mine Type	Average Temp Differences (°C) between loose and solid rock	Average air-to-rock temp differences (°C)
Coal	0.2 to 0.4	1.2
Zinc	.3 to .5	.9
Copper	.3 to .5	.7
Lead	.2 to .5	1.0
Building stone	.3 to .5	.8
Others	.2 to .4	.8
Total non-coal	.3 to .5	.8

Local mine environments have a significant effect on loose rock detection with infrared techniques. In Cominco's Sullivan mine, Kimberly, British Columbia, very few temperature differences between loose and solid rock were detected, even when the rock was visibly loose, because no significant air-to-rock temperature contrast existed. Consequently, conditions in the areas examined were very unfavorable for loose-rock detection by temperature differences. In contrast, conditions were more favorable in the Kerford and Colony mines. Temperature differences were easily detected throughout these mines and were not restricted to newly blasted faces. This abundance of temperature differences between loose and solid rock was the direct result of favorable contrast in temperature between air and the rock surface.

Coal mines also presented conditions that varied in favorability for loose rock detection by infrared techniques. Unfavorable conditions existed in most areas of the mines visited in Pennsylvania and West Virginia. For example, all these mines were using continuous-type cutting machines with a continuous spray of water. The water tended to cause the temperature over the whole face to be in equilibrium. Often the equipment was not out of the opening long enough for temperature differences to develop. Also, air tubes used in ventilation of the mines created temperature gradients across the roof of the new rooms so that it was often difficult to identify temperature differences between loose and solid material. However, temperature differences could be detected when access to the face was available and temperature gradients were allowed to subside.

By contrast, conditions in the coal mines visited in Alabama were more favorable. In these mines, conventional mining methods (drilling and blasting) were used. After shooting, each entry was washed down with water to settle the dust. Because a longer time elapsed before mining resumed in the room, not only was the effect of the water nullified, but probably evaporation caused more rapid development of temperature differences between loose and solid rock. Also, brattices were used to control the flow of ventilation air. This practice provided a more even air flow and eliminated the temperature gradients that had been caused by air flow through ventilation tubes.

Another factor that influences the use of infrared techniques in coal mines is the fact that coal has a very low thermal conductivity. Consequently, when "top coal"** is left on the roof of the mine, temperature differences develop more slowly and may be smaller. As a result, loose material is more difficult to detect.

It should be noted that in all large coal mines, temperature differences between loose and solid rock have been very small, or nonexistent, in the main haulageways. Apparently, the rock surface temperature in these areas remains near equilibrium with the air temperature.

Figures 1 and 2, show a visual and a thermal image of loose material on the rib of a coal mine in New Mexico.

Fig. 1. Visual image of loose material on the rib of a coal mine.

**Many mines leave a layer of coal on the mine roof in an effort to decrease the danger of "roof falls."

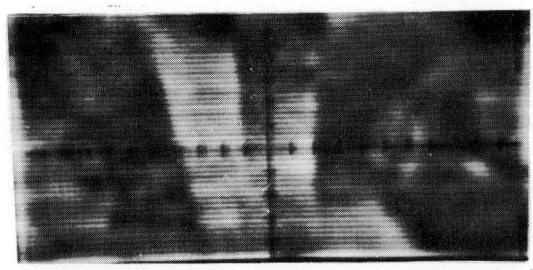

Fig. 2. Thermal image of loose material on the rib of a coal mine.

Misfire Detection

The concept of using infrared techniques for detection of misfires was first discovered in the Kerford mine, Weeping Water, Nebr., when small "hot spots" were noted on a new face following blasting. These hot spots were from 1^0 to 1.5^0C warmer than the surrounding rock surface for as long as 3.5 hours after blasting. The "hot spots" resulted from the residual heat around the ends of shotholes on the face that had been exposed by the blast.

Further tests with both the infrared thermometer and the scanner, in the Kerford mine and in the Colony mine near Rifle, Colo., indicated these shothole locations could be detected by infrared techniques 8 hours or more after blasting, that they could be as much as 5^0C warmer than the surrounding rock surface, and that the position of the shothole in the pattern could be determined. Therefore, the absence of a hot spot representing a shothole in a known pattern indicates a misfire.

Electrical Faults

The detection of faulty electrical equipment with infrared equipment has been used effectively for several years by power companies. Until recently, most of the equipment so used was truck mounted and operated by three-man crews. Obviously, this type of equipment is not practical in mine openings, especially those with a vertical dimension measured in inches. Therefore, portable infrared imagers were evaluated for this type of application.

A feasibility test was made in the laboratory by shorting a section of cable normally used as power cable by coal mining machinery. Although the voltage was much lower than that used in coal mining, a hot spot was quickly generated in the cable and was easily detectable with infrared imagers. The ultimate test occurred in the SEGCO No. 2 mine near Maylene, Ala. When the investigators arrived at the mine, it was learned that one of the work crews had been working for about 4 hours to locate a short in a trailing cable (power cable). By using an infrared imager, the short was located in about 10 minutes. The surface of the cable over the short was more than 1^0C warmer than cable surface a short distance away. Undoubtedly, the thermal anomaly was much greater when the short first occurred.

Hot Bearings on Conveyor Belts

Infrared imagers have been evaluated for detection of hot idlers (bearings) on conveyor belts. Beltways were examined in 5 mines. Excessively hot bearings were found in three mines. In the other two mines, several bearings were found that were warmer than usual but were not hot enough to indicate approaching failure. However, when cleaned and oiled, these bearings ran at approximately the same temperature as adjacent idlers on the beltway.

During the course of this investigation, the following was determined:
1. The imagers could not only detect overheated bearings, but could detect them before serious problems developed.
2. Infrared techniques could be used to quickly examine large systems of conveyor belts. The operator can examine from 100 to 200 feet of belt from each vantage point. Consequently, the time required to move from one point to another is the chief limiting factor. In one mine, one operator examined 5 miles of beltway in less than 4 hours by using an electric cart for transportation.

Figure 3 is a photograph of the imager display showing a hot bearing on a conveyor beltway. The very bright spot in the foreground is a hot bearing. Other bearings, the belt loaded with coal, and even the hangers supporting the belt can be identified.

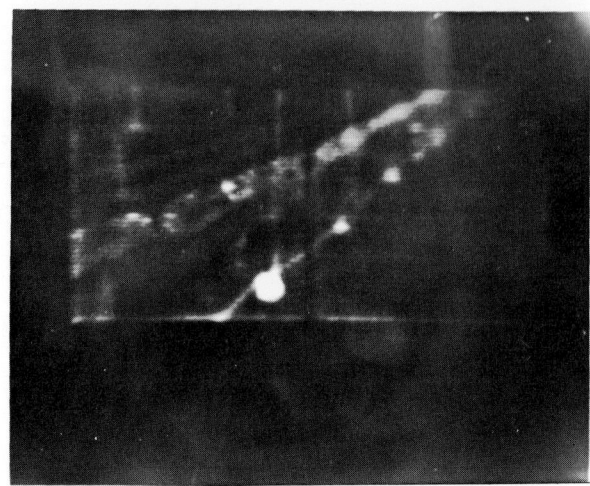

Fig. 3. Thermal image of a hot bearing on a conveyor belt.

Fires in Dumps and Outcrops

Another mine-related application of infrared imagers is the detection of fires in dumps and coal beds at their outcrop. Fires of this type can be a serious hazard to life, environment, and property. For example, these buried fires emit noxious gases, kill vegetation, and permit percolation of acidic water. Airborne application of infrared instruments has been demonstrated to be effective in the detection of dump fires but has been less successful in the detection of fires in place in the coalbeds. In addition, overflights with airborne equipment can be expensive. Handheld instruments were evaluated for this purpose because they could be carried in any vehicle and could be used repeatedly at small expense once a fire was located. As a result, the progress of advance of the fire or efforts to put it out could be monitored from remote locations.

Eleven coalbed fires were examined during this study. Ten of these were known to exist prior to the investigation, and one was discovered during the study. Although no maximum range was determined during the study, the one unknown fire was detected from a vantage point about 1-1/2 miles away. The temperature differences between the surface over the fires and the surface of virgin ground ranged from 8^0 to $40^0 C$.

More than 100 coal mine dumps have been examined with infrared imagers. Thirteen of these were found to have buried fires at the time they were examined. Surface temperature differences on these dumps ranged from 10^0 to $55^0 C$. The most interesting of these investigations were made by using the portable imagers from small aircraft. An area of more than 800 square miles was examined in about 2-1/2 hours. Fires were detected in several dumps at ranges of more than 2 miles. Confirmation on these fires was made by infrared investigations on the ground at the dumps. Three of the dumps that the airborne study indicated to be clear of fire were examined from the surface on and around the dumps. The results here also agreed with the airborne study.

In addition to coal mine dumps, eight dumps of flood-generated trash were examined near Wilkes-Barre, Pa. These dumps were old strip pits that were filled with trash and debris after Hurricane Agnes flood that hit part of the country. Three of these dumps had thermal anomalies on the surface. The magnitude of these anomalies on the surface was about 3^0 to $5^0 C$ but at a depth of 12 inches below the surface they were as great as $50^0 C$. All three of these dumps were openly burning within 6 months after the study [3] and contaminating nearby residential areas with noxious odors and gases.

Incipient Mine Fires

Spontaneous combustion in coal mines can be a very serious problem, especially those mines where residual heat is present in the coal and/or conditions favorable for oxidation. Usually fires resulting from spontaneous combustion are monitored by measuring the carbon monoxide (CO) content of the air leaving the mine. This method detects the existence of the fire but does not locate it. Infrared imagers offer a means of locating temperature increase before combustion begins. Figure 4 is a thermal image of a hot spot on the surface of a pillar in a coal mine. This hot spot has a minimum diameter of about

Fig. 4. Thermal image of a hot spot on a coal mine rib.

6 feet. The temperature in the center of the spot is 60^0C as compared with 32^0C on the surrounding surface. Spontaneous combustion in the coal at this location tends to be an irreversable process once it reaches a temperature of about 50^0C, and seems reasonable to conclude that in this case, at least infrared techniques have detected an incipient mine fire[5].

Mine Rescue

Another potential application of infrared techniques in mines is the ancillary use in the rescue of men from hazardous conditions. Rescue teams that enter mines after fires or explosions normally encounter dense smoke that reduces visibility to near zero. The objects sought by the rescue teams (whether men or the fire itself) are hidden by the smoke. Infrared imagers have been evaluated as tools to see through the smoke[6].

An opening in the Bruceton Experimental mine near Pittsburgh, pa., was filled with smoke so dense that a mine lamp could not be visually detected at a distance of 5 meters. Two men, wearing breathing devices, entered the smoke to serve as targets. They were observed at 5 meter intervals over distances from 5 to 75 meters. At 75 meters, the targets were identifiable as men although the resolution was reduced.

Figure 5 is an excellent example of the effectiveness of infrared imagers for smoke vision purposes. The instrument providing this image was originally designed to detect men at 50 meters. The men in this photo are 50 meters from the instrument in the dense smoke.

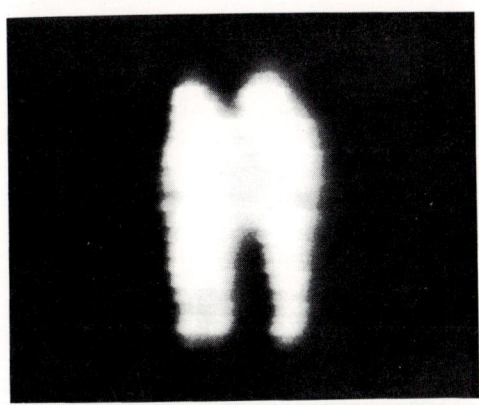

Fig. 5. Thermal image of men in smoke

Because of the favorable results of the smoke vision tests, calculations were made to determine probable effectiveness in various smoke densities for instruments operating in the 3- to 5μm region. Figure 6 is a graphic representation of the results of those

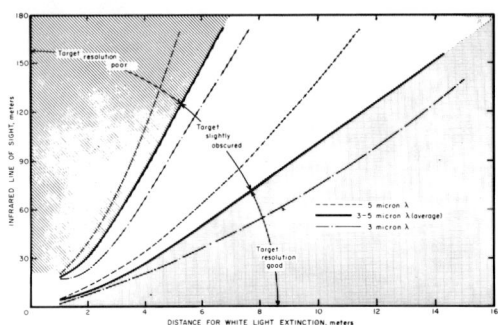

Fig. 6. Infrared smoke vision capability chart.

calculations where smoke densities are differentiated on the basis of a white-light extinction factor. In mine rescue applications, this factor would be determined by the distance at which a miner's cap lamp becomes invisible in the smoke. To use the graph, the light extinction factor is read on the horizontal scale. One can project up to the first solid line, thence back to the left to the vertical scale, and directly read the distance over which good resolution can be expected. Projection upward to the second solid line, thence back to the left determines distance over which one would have fair resolution. Beyond these distances, the targets would be detected, but resolution would be doubtful.

Instruments used in these studies were not designed to have good image resolution past 50 to 75 meters. Examination of the graph indicates the design factor is a greater limitation than is the smoke. For example, the light extinction factor for the smoke used in the studies was 5 meters. Predicted good resolution in this smoke is about 40 meters with fair resolution for targets more than 120 meters away.

Conclusions

The results of these investigations indicate that many different types of mine related hazards can be detected with portable infrared imagers. These hazards include loose rock, misfires, short circuits, potential and existing combustion in mines, dumps, and coal beds, and hot bearings on the belt line. In addition, the same infrared imagers can serve as useful mine rescue tools by providing a way to see through dense smoke.

References

1. Merrill, R. H., and R. M. Stateham. Hazards Can Be Detected by Infrared Remote Sensing Techniques. Proc. Internat. Strata Control Conf. National Coal Board, London, 1972, pp. 21-25.
2. Merrill, R. H., and R. M. Stateham. Loose Rock Can Be Detected by Infrared Devices. Min. Eng., v.22, No. 11, November 1970, pp. 50-62.
3. Stateham, R. M. Detecting Hot Areas in Dumps With a Handheld Infrared Scanner. BuMines TPR 68, 1973, 12 pp.
4. Stateham, R. M. Field Studies on an Unsupported Roof, York Canyon Mine, Raton, N. Mex. BuMines RI 7886, 1974, 19 pp.
5. Stateham, R. M. Remote Sensing of Potential Spontaneous Combustion in Coal Mines. BuMines TPR 80, 1973, 9 pp.
6. Stateham, R. M., and M. L. Bowser. Using Handheld Infrared Imagers To Locate Miners in Smoke-filled Mine Openings. BuMines TPR 87, 1975, 11 pp.
7. Stateham, R. M., and R. H. Merrill. Detection of Hazardous Conditions in Mine Operations through Infrared Techniques. Conf. Underground Mining Environment, Univ. Missouri--Rolla, Rolla, Mo., October 1971, pp. 53-67.

INFRA-RED IS NOT A PANACEA - USE DISCRETION

Alan J. Van den Berg
Facilities Engineering Support Agency
US Army Corps of Engineers
Fort Belvoir, Virginia 22060

Abstract

Infra-red thermography by itself, despite much wishful thinking, will not quantify heat losses, nor is it always cost-effective when used for other purposes, but it does have a place as a useful tool when used with discretion.

Deep down inside all of us must be the spirit of an alchemist. Most of us resist the temptation to defy the laws of nature. But some, having lived in the so-called atomic age, in which gold has, indeed, been synthesized, feel that somehow, somewhere, they too will be able to repeal a previously immutable law of nature. There are those, however, who knowingly give false advertisement, as did some of the ancient alchemists, that they have the solution in hand. They then proceed to dupe the public as the alchemists cheated the kings of old who subsidized the elusive conversion of brass to gold.

We were all taught early in our lessons in mathematics that you cannot solve an equation having three dependent variables when you know the value of only one variable. Yet daily we hear people saying that they can use infra-red thermography to determine the heat losses from a building.

Infra-red devices, whether exotic, airplane mounted photographic scanners or prosaic hand-held thermal radiation probes are basically radiation temperature dependent devices - they indicate the apparent temperature differences on the surface at which they are aimed or they may indicate the delta T between the surface and a built-in standard. Temperature however, is only one variable in the heat transfer equation. If radiation losses are to be considered, one needs to know the emissivity of the radiating surface. If convective heat losses are to be measured, then the conductivity of the wall must be known. Rarely is the conductivity or emissivity of the total structure accurately known and, hence, you cannot use temperatures by themselves, no matter how accurately they may be measured nor how cleverly they may be displayed, to determine heat losses. No one seriously contends that you can ascertan the power used in an electric circuit containing an unknown impedance by simply measuring the voltage. Yet some think you can solve the analagous problem of heat transfer by merely measuring the temperature. Contrary to some people's expectations, changing an indicating D'Arsonval meter to a digital display or substituting a rainbow of discrete colors for shades of gray from black to white, does not, through some exotic means, supply the missing unknowns in the heat transfer equations and miraculously provide one with a quantitative value of heat losses. I must admit that all those pretty colors help sell a consultant's final report to an unwary customer, but to date, I know of no one who has been able to quantify the heat losses from a building using only infrared devices. Conceivably, some combination of instruments will be found that will give the required results. The Corps of Engineers has been working with heat flux-meters in combination with infra-red devices to try to find a quick way to quantify heat losses. As yet, however, we have no heat flux meters that don't have to be attached to the surface being measured.

There is a disturbing question in the background of all of this work. What are you going to do about it if, eventually, you can readily quantify the heat loss from an existing building. Saving energy may be the patriotic thing to do but, in the real world, cost-effectiveness must be considered. At a building on Eilson Air Force Base, Alaska, infra-red photographs were taken and the qualitative decision was made that the building was losing heat through the walls and roof and through infiltration around the windows. Using conventional heat transfer methods -- flux meters, thermometers, and the assumption of insulation and other construction material "U" factors, etc., it was determined that about 60 tons of coal could be saved per year if new ceiling insulation was installed. At the existing price of coal to the Government in Alaska that amounted to about $950/year saving. Fixing the walls would save another $910/year and installing storm sash windows another $380/year. The total saving would amount to $2240/year -- representing more than 130 tons of coal per year for that one building alone. Cost effective? No way! The investment cost of the renovation would amount to $265,400. The price of coal would have to escalate to over $200/ton to be able to amortize that capital investment in 30 years -- and the building is already 35 years old.

I realize that good sense doesn't always come out on top when the glamour of something new or different is proposed. Recently an East Coast Army Depot was overflown with a line scanning infra-red camera. The operation was highly acclaimed for the great "heat loss" it had found. For instance, open windows, roof ventilators, and back doors ajar were discovered -- things the night security guards could have told them at no cost, if they had been asked. The investigators found a condensate pump that was discharging into the sewer instead of returning the water to the power plant. Why the plant operators hadn't looked for trouble when the boiler make-up demand rose sharply was not explained. The most "astonishing and worthwhile" thing they found was a 500 gallon hot water heater operating in an unused mess hall. But what they didn't say was that the boiler was well insulated, and since no hot water was being taken from it, it probably didn't require fifty cents worth of steam a day to make up for its insulation losses. At that rate it will take many years to amortize the cost of the flight and the photographer assuming, of course, that no one would have turned off the boiler in that time. The commanding officer was impressed, however, and ordered all similar installations under his command overflown with infra-red thermographic equipment immediately!

Where qualitative answers are satisfactory or where temperature differences in themselves are significant indicators of trouble, infra-red devices make excellent diagnostic aids. One highly remunerative use of infra-red scanners and probes is in connection with the periodic inspection of electric transmission and distribution lines and associated equipment. Loose connections, overloaded transformers, defective high-voltage switchgear, and other similar problems are readily detected with an infra-red scanner. Most importantly, the entire inspection usually can be accomplished without shutting down any of the power lines. Reuently at Fort Gordon, Georgia, a complete inspection was made of a combination transmission and distribution sub-station in less than an hour. And this included time to take unnecessary photographs of some of the infra-red displays to prove we had been there. This task would have normally taken from 40 to 50 hours of downtime - on Saturdays and Sundays - with all the resultant problems that shutting down the power supply on a big post entails. The overtime labor charge is enough to make one shudder. In addition, since all the connections would have to be mechanically torqued, damage certainly would have occurred when somebody pulled a wrench too hard and twisted off a nut on a copper bolt or strip the threads on a terminal board screw. An infra-red scanner seems to be obviously desirable. But this is a "worst case" comparison and is not a valid comparison of competing alternatives. The same sub-station was inspected a year ago using a hand-held radiation temperature probe. The task took about eight hours to accomplish. You cannot, however, shrug off the probe in favor of the scanner without considering the capital costs of the two instruments. The scanner cost 40 times the price of the probe and, unless it is used for a lot more tasks than an occasional sub-station or motor-control center inspection, it would not be the most cost-effective tool to do the work. It takes a lot of productive usage to cost-effectively amortize 40 thousand dollars worth of instrumentation.

One use of infra-red scanning which can provide an immediate pay-off is in the detection of moisture in the insulation of a flat, built-up roof. Here, as in the detection of pregnancy, there is no quantitative question involved -- there is a "yes" or "no" answer. "How" leaky a roof is, is not the question -- "Where" are the leaks, regardless of amount, is the pertinent factor. Here the infra-red scanner is most effective. Even though it cannot be used in the daytime because of signal to noise problems caused by the sun, it can be used in either the summer or winter. In the winter, the heat leaking through a roof from the interior of the building selectively raises the temperature over the more highly conductive wet insulation and the wet spots under the roofing felts can be seen easily in a hand-held scanner. In the summer, the sun's heat raises the roof temperature during the day. At night the wet insulation acts as a heat sink and the surface temperature over the defective insulation is higher than the surrounding roof. Is it cost effective? Where large built-up roofs -- acres in extent -- are to be examined a scanner can pay for itself in a single night. For example, on one condemned roof in Alaska we determined that there were only two spots, each about 15 by 30 feet in extent, that were defective. Including the cost of termination of the reroofing contract and the cost of repairs, the saving that night amounted to over a quarter of a million dollars. On other roofs we have found small leaks that had just begun to destroy the insulation. Early detection and pin-point location of defects can save large amounts of money later on. Is aerial surveying effective for this work? Here again, the number and extent of the roofs must be considered. All aerial scanning will do is to point out places that are candidates for hand-held scanner inspection. From the air it is not possible to tell the difference between water on the surface of the roof and water under the felts. Nor can differences in radiation temperatures due to different surface emissivities be accounted for. Even when walking a roof with a hand-held device, the answers concerning temperature differences are not always immediately apparent. Varying thickness of bitumen have caused some erratic results. The presence of heated devices, pipes and ducts, for example, have also given preliminary false indications on the scanner. However, when used with discretion, the infra-red scanner is by far the best way to inspect built-up roofs. And bad built-up roofs are no small business - this year the Army has over 50 million dollars worth of backlogged roofing maintenance and repair.

Infra-red thermography is great for determining the pattern of power plant hot-water effluent in a stream. It is great for detecting changes in agricultural growth. It works wonders in finding enemy troops in the jungle and it has no substitute on fighter planes flying at night under visual control. It can be used to find steam and hot-water lines underground. You can also use it to find out if someone left the back door open. But it will not tell you how much energy is being lost nor will it pinpoint where a pressurized hot-water line has broken in an underground conduit. And if an unconduited high-pressure steam or water line has broken, you don't need an infra-red instrument to find the leak.

Infra-red thermography has its place but it is not a panacea for all the problems of our energy conscious public. However, when used with discretion, and with full understanding of its shortcomings, it can be a very useful tool.

The Technology Behind X-Ray Security Systems

David J. Haas
Philips Electronic Instruments
Mount Vernon, New York 10550

I   The History of X-Ray Technology as Applied to Security Screening Devices.

Within a year after x-rays were discovered by Roentgen, x-ray imaging was being used not only for medical applications, but also by Customs and Security people for examining the contents of packages and parcels. The full potential of the penetrating nature of x-rays with its ability to produce easily recognized images was quickly recognized, but its implication was impractical and proved less valuable for security use than for use in the medical profession. Hence, x-ray security systems faded away and were employed only on a very limited basis for about 70 years (1900-1970). During this period, manufacturers were offering industrial or medical x-ray units primarily designed for other functions. These machines failed to develop a viable market and no more than a few dozen units were actually employed in security functions. Even though the machines presented an x-ray image of the parcel, why did security officers avoid them?

Well, there are multiple reasons. They all looked like industrial equipment and were unsuited for offices, public buiildings, etc. They were big and heavy, not really portable, and presented the operator or guard with a rather dim image in an eye-viewer. Furthermore, the x-ray chamber door through which items were loaded was generally very heavy and slow to operate. All of these parameters are related to the high x-ray dose-rate required to form the visible image on the fluorescent screen (the transducer of x-ray photons to visible light photons).

These units are called today "high dose" because the light output is generated only by the large x-ray dose impinging on the fluorescent screen. In addition, high kilovoltage was employed in order to generate the large x-ray flux from the x-ray tube. One factor not directly related to the hardware itself is the effect that large x-ray exposures have on photographic film that might be in the item being examined. High dose units will create shadow images on unprocessed film, so that is a security application where film may be present, special precautions are required. If hand-searching of the parcel is necessary in a high percentage of the cases, the value of the x-ray screening device is reduced and may even become a disadvantage because of security film damage.

II   The Technology Behind the Recent Success of X-Ray Screening Devices.

X-Ray machines became practical for security applications within the last 5 years. The main reason for this was the ability to generate an x-ray image using low levels of x-rays, so low in fact, that they would not even damage photograpic film. A low-dose unit employs an x-ray dose rate (flux) which is less than 50,000 times that of a high dose system. This enormous reduction in x-ray dose-rate exhibits itself in the x-ray system as:

     1.   reduced lead shielding for radiation protection
     2.   lower x-ray tube kilovoltage
     3.   lower power x-ray generator
     4.   the ability to have an easy passage into the x-ray chamber for loading and unloading items to be examined.

Independent of all other factors, this is directly due to the ability to operate the x-ray system with low x-ray dose-rates. As the radiation safety requirements are that the leakage radiation from the x-ray machine be less than 0.5 mR/hr at 5 cm from any surface, the low x-ray dose-rate of the primary beam means a proportional reduction in the scattered radiation. Thus, high dose systems operate with primary beam fluxes greater than 1000 R/hr with scatter in the x-ray chamber of 1-10 R/hr while low dose units work at less than 1 R/hr so that the scattered radiation from the x-ray chamber is of the order of 1-10 mR/hr, the same order of magnitude of the permissible leakage level.

One additional factor is the requirement for a bright display image.

Security systems are used in well lighted locations by a large variety of personnel. High dose units have traditionally offered rather dim images directly from the fluorescent screen which requires the operator to view the image with a light-tight eye shield. This is objectionable and impractical for security systems.

Thus, low dose x-ray units were developed to satisfy the airport screening function and have proved acceptable to the security industry. Four types of designs are currently being employed in low-dose systems:
1. pulsed x-ray source with TV image storage display;
2. continuous x-ray beam with realtime TV image display and,
3. continuous x-ray beam with realtime direct viewing display,
4. flying-spot scanner type.

Each system has its related advantages, but two factors have become pre-eminent in these security x-ray machines: 1) system cost and 2) equipment reliability.

Each of the four system designs require an x-ray flux about 50,000 lower than that of a high dose unit. In all the systems except the flying spot scanner, this is accomplished by light amplifiers, generally two to five stage units.

The pulse systems work as follows: An x-ray generator gives an intense pulse of x-rays, usually less than 100 milliseconds long. The burst of x-rays passes through the item being inspected and impinges on the fluorescent screen. The x-ray photons are converted into light by the fluorescent screen and a visible image appears for about 10 times longer than the x-ray pulse length (this is due to the long decay time of the fluorescent screen phosphors). A television camera is synchronized with the visible image emissions and the picture is captured and stored on a video storage device. Then the image is displayed to the operator until the next inspection cycle begins.

The second system is the continuous beam/TV imaging system. Here the x-rays strike the fluorescent screen while the item being inspected is in the x-ray chamber. The TV monitor displays the real time fluorescent screen image so no storage device is required. The third system is identical to this except no television camera or monitor is employed but light intensifiers are used alone with lenses or enlarging display tubes.

Let me now describe the flying spot scanner type system for it is by far the most complicated device. A continuously operating x-ray tube has a mechanical collimating device which produces a horizontal pencil beam of x-rays that sweeps up and down along a vertical path. The item being inspected transverses the x-ray beam on a horizontal conveyor belt, so that the x-ray beam sweeps up and down the entire length of the item as it moves along on the conveyor belt. Instead of a two-dimensional fluorescent screen transducer, a long vertical scintillation crystal is used to convert the x-ray photon to visible light photons and an attached photomultiplier converts the visible light level into electrical signals. These signals are stored in a video storage device in a two-dimensional array which, upon completion of the scan, is displayed on a television monitor. Thus, the x-ray image is formed sequentially by the flying spot scanner while all other systems form the x-ray image simultaneously on a fluorescent screen.

Needless to say, the four approaches to low dose security employ a wide variety of electrical, optical and mechanical devices, but the market place will ultimately determine which are to remain competitive and viable products.

Rather than spending extra time on discussing each technology, I would rather summarize the advantages and disadvantages of each approach. This is simply because the security market has already selected between the various technologies and is forcing manufacturers to produce only selected types of systems.

In the previous section I pointed out that the security industry required low cost, reliable units. For this reason, the market is selecting the technology that is least expensive and simple, independent of some other outstanding advantages of the more sophisticated technologies. For example, the flying spot scanner system gives an x-ray dose to an item being inspected from one-tenth to one hundreth of any other system. Yet it is the most expensive and most complicated type of device. The pulse type units have the advantages of being light weight, but they too suffer from complexity and cost. So from my vantage point today, I believe that only continuous beam systems will survive, and probably only those without television because the cameras required are expensive low light level cameras.

In order to clarify this last point, only special low light level cameras can be employed with x-ray systems because the useful fluorescent screens emit in the green/yellow region (the peak is 5400A) with essentially no red or blue emission. So SIT, I SIT and silicon diode pickup tubes generally cannot be employed.

III The Pragmatic/Economic Aspects of X-Ray Security Systems

Even though this is a technical paper, I feel it is essential that I inform the nonsecurity people about the practical aspects of the x-ray security business. Security is an overhead expense and a security officer cannot always demonstrate direct dollar savings by his expenditures. Thus, many times product cost eliminates desirable technologies because of the economic justification.

The airport x-ray screening systems are run 15-20 hours a day by unskilled "guards" (male and female) from ages 18 and up. They are often serviced by local ground maintenance personnel who have never had any training in x-rays. So, traditionally, industrial x-ray manufacturers had to provide systems to an entirely new type of operator so that no mistake could be made during its operations.

Figures 1 and 2 show a continuous beam Direct Viewing Conveyorized system now in service at airports.

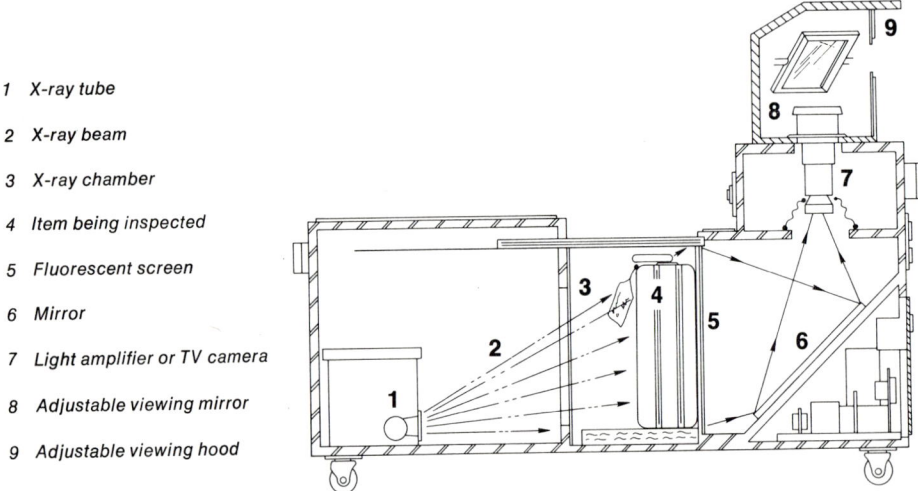

1 X-ray tube
2 X-ray beam
3 X-ray chamber
4 Item being inspected
5 Fluorescent screen
6 Mirror
7 Light amplifier or TV camera
8 Adjustable viewing mirror
9 Adjustable viewing hood

Figure 1
Pictorial View of Direct Viewing Security System

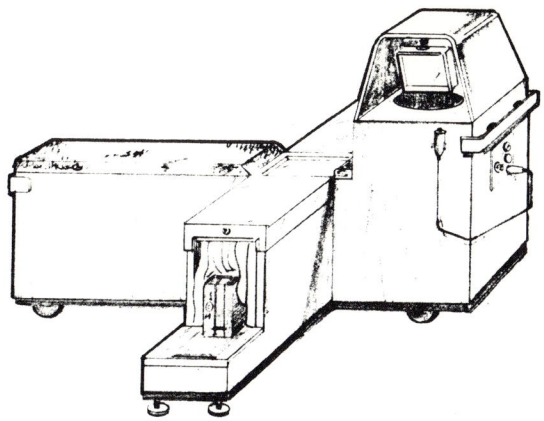

Figure 2
Configuration of Conveyorized System

The operator control panel has only a power on key switch, an emergency off reset button and a conveyor off switch. A remote handswitch permits the operator to start and stop the Conveyor/X-Ray System. No meter or service controls are accessible to the operator and all service panels have to be key-locked because during slow periods the operators tend to "tinker" with the equipment.

The radiation protection of the machine is permanently attached because these machines receive considerable abuse. The only service adjustment on the entire optical system is the objective lense. Operators have a mirror for adjusting the viewer to their convenience.

IV   Future Technological Advances

The future developments that can be envisioned are:

1) improved fluorescent screens with greater light output of a more suitable spectrum for current light amplifying devices.

2) light amplifiers sensitive to the current fluorescent screen spectral emission.

3) light amplifiers with superior resolution.

4) display tubes or lenses to enlarge the intensifier field of 25 or 40 mm for easier operator viewing and larger field presentation.

5) Some other type of x-ray to light transducer which would replace the current fluorescent screen/image intensifiers combination.

6) superior storage system for pulsed x-ray designs.

All the above items must, of course, be lower cost than existing items since sales of the security x-ray units are very price sensitive. The x-ray portions of security systems have already shown remarkable development, so I feel the next advance can only be in the imaging systems.

As a last word, I want you to appreciate the relative cost of security systems components. The imaging system represents about one-half of the entire system cost, while the cabinets, x-ray systems, electronics and transport devices constitute the other half. So real cost savings can be appreciated by reducing the imaging component cost.

USE OF NIGHT VISION SYSTEMS BY THE LAND MANAGER

Herbert J. Shields
Project Leader, Equipment Development Center
USDA Forest Service
San Dimas, California 91773

Abstract

For the last two years, the USDA Forest Service has been engaged in an R&D program to expand helicopter firefighting operations into the nighttime by use of current night vision technology. As a result, Land Managers are beginning to utilize some of the systems and devices for other tasks as well. These include law enforcement on National Forests, surveying techniques, nocturnal game studies, search and rescue, and reconnaissance duties. This paper describes the equipment in present use, training requirements, and typical operations.

Introduction

In 1973, Congress appropriated special funds to the USDA Forest Service to investigate new potential techniques which would help alleviate the severe wildland fire threat that exists nationwide. One such project initiated was designated "Helicopter Night Operations." Several other agencies participated in this investigation; the Los Angeles County Fire Department (LACFD), California Division of Forestry, Bureau of Land Management, Oregon Department of Forestry, Aerospace Corporation, U.S. Army, and U.S. Air Force. The exploratory capability was achieved through the use of image intensification devices and thermal imaging systems installed on a UH-1M helicopter loaned by the U.S. Army's Night Vision Laboratory.

Fig. 1. UH-1M equipped with Aerojet AAQ-5 FLIR.

One of the Land Manager's major weapons against unprescribed wildfires has been aerial support in the form of chemicals and water for either direct suppression or line-building. This support is enhanced at night because of depressed temperatures, decreased winds, and elevated humidities.

Systems

Our first activity in the project was to become familar with current hardware that would: (1) be available to civilian users as unclassified equipment, and (2) hopefully fit the environment of Federal and State land primarily consisting of rugged forested terrain. Our original plans called for evaluation of an INFANT low-light-level (LLL) system and a FLIR (AAQ-5) system; both loaned from the Army. However, a chance exposure to the availability of the PVS-5 night vision goggles (NVG) quickly preempted further consideration of INFANT trials. Six pair of NVG were purchased for the project, one pair of which was equipped with bifocal objective lenses. Having a field-of-view of 40 degrees and weighing less than 2 lb, the NVG can be worn on the head by attachment to a flight helmet, thus permitting a hands-free operation.

After initial training and test flights were underway, a pink-filtered light was loaned by Spectrolab, Inc., Sylmar, Calif. for trials as a supplemental illumination for subthreshold takeoff

Fig. 2. Night vision goggles worn by pilot.

Fig. 3. IR floodlight.

and landing areas. This light utilizes a 1 kW Xenon lamp, and projects a wide-angle flood (180 degrees) having a vertical field of 60 degrees. The benefits of this type of diffused lighting when landing in a dark canyon bottom are enhanced vision when most needed; also persons in the area are unaware that it is being used. This tends to keep well-meaning people from illuminating bright lights in the landing area, which can disturb the pilots.

Although the FLIR was used for display purposes and occasional intelligence, it became obvious that when operating in mountainous terrain, a monitor-type display was unsuitable and unsafe when used as the primary reference for flight. However, it provided a redundant scene which did prove useful navigationally when used in conjunction with other aids.

One additional system also evaluated was the TALAR portable instrument landing system (ILS) manufactured by the Singer Company, Kearfott Division. The equipment consists of a ground transmitter which radiates microwave guidance beams, and an airborne receiver which interprets the information and provides localizer and glide-slope signals similar to conventional ILS. These signals are presented in a standard cross-pointer instrument display. During the initial experimental work, the FLIR image provided the second pilot with a visual proof of position over the ground as well as key checkpoints, while the first pilot was flying the ILS approach.

Recent acquisition of a MINI-FLIR, manufactured by Aeronutronic-Ford, Newport Beach, Calif. has provided an expanded role for this kind of equipment. The primary improvements of this second generation FLIR have been a drastic reduction in weight and size, and display imagery having a video interface in 525 line standard TV. This permits recording imagery on a television videocorder, as well as dubbing in additional information of time and date. Thus, the tape can be replayed after the flight mission, and retained for a permanent record. The basic data regarding this system are as follows:

FLIR Sensor:
 Type: Two row serial scan
 Spectral region: 8-11.5 microns
 Detectors: HgCdTe
 Detector IFOV: 2mR (nominal resolution)
 Scanned FOV: 30 x 40 degrees
 Noise equivalent temperature: Less than 0.25°C

Overall System:
 Power: 11A at 24 to 30 VDC
 Weight: Basic FLIR system—73 lb
 Weight accessory package: Two TV monitors; power supply for videocorder; videocorder; time-date generator; assembly frame for helicopter installation—50 lb.

Fig. 4. Portable ILS ground station.

The installation of the FLIR system was designed and constructed to permit reasonably fast mounting and demounting of the two subsystems. After developing some experience, 10 minutes is the normal installation time; and an additional 10 to 15 minutes is required to obtain cool-down for proper detector operating temperatures. The only required special servicing equipment for routine field operations, is a small pump for maintaining proper vacuum in the scanner.

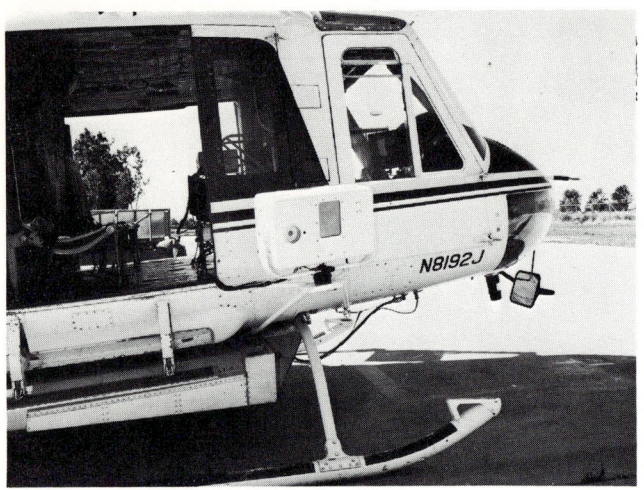

Fig. 5. Aeronutronic FLIR sensor.

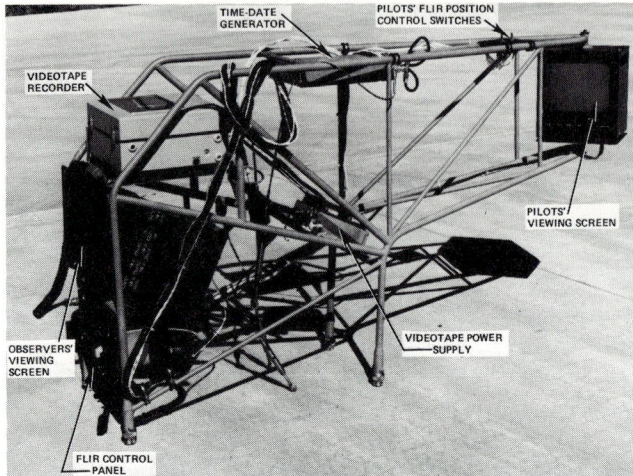

Fig. 6. FLIR accessory package.

Several basic features are essential in most night vision equipment to be utilized in civilian activities:

Easy installation and quick removal— Major aircraft modifications may require extensive certification procedures with the FAA.

Interchangeable between helicopters— If either the aircraft or equipment is down for repairs, the working system should not be compromised by reason of being permanently attached. Also, equipment can frequently be used on other aircraft operating in a different mission scenario.

Low maintenance, high reliability— Civilian operations must be cost-effective. Availability and reliability must be high during field operations when manpower is taxed for normal aircraft service. If the equipment doesn't do the job intended, other alternatives are selected.

Training

Before long, the user of night vision equipment begins to wonder what the limitations of the equipment and environment are for his task at hand. And, since this new technology is being introduced to people schooled in other disciplines, a new language must first be learned. A typical outline for training in essentials for occasional users is as follows:

SUBJECT MATTER OUTLINE

I. OPTICAL SPECTRUM
   A. Visible
   B. Infrared
   C. Low-light level (LLL)

II. INFRARED
   A. Principles
   B. Parameters of Performance

III. LOW-LIGHT-LEVEL SYSTEMS
   A. Principles
   B. Image Intensifiers

IV. LIGHTING
   A. Light-level definitions
   B. Supplemental

V. NIGHT OPERATIONS SYSTEMS
   A. LLL devices
   B. Infrared devices
   C. Other supplemental aids

After helicopter pilots gain experience in the use of LLL devices the ultimate key to safe use is understanding light levels in the night environment. This includes knowing when threshold (lower limits) are being encountered, as well as how best to conduct operations when existing light is uncomfortably intense.

After having worked with a population of 13 pilots, the initial requirement was evident that they be current and experienced in the machine to be flown as well as the environment in which it will be operated. As aptly stated by one, "There is too much going on when flying with the goggles to worry about flying the helicopter." In addition, it is imperative that the machine be of the governed-turbine type. At the present state-of-the-art, two pilots should always be utilized for flying with NVG, since there is always the possibility of equipment failure. A syllabus developed for pilot training includes the following:

Pilot Training in the use of NVG
LOW-LIGHT-LEVEL (LLL) AMPLIFICATION DEVICES ELEMENTARY THEORY AND FLIGHT TRAINING

- FIRST HOUR—*High Light Level*—Primarily establishes capability to hover and perform takeoffs and landings under ideal conditions.

- SECOND HOUR—*Medium-to-High Light Level*—Expands flight into mountainous and canyon terrain to work with varying light levels. This develops new visual cues.

- THIRD HOUR—*Low-to-Medium Light Level*—As light level diminish, pilot works on ability to fly by "seat-of-pants" and to judge light level and resolution limitations.

- FOURTH HOUR—*Low Light Level*—Two pilot team work and limitations of "no-go" are established at this point of training.

- FIFTH THROUGH TENTH HOURS—Proficiency and confidence is established by practice under as many field conditions as possible.

- POSTGRADUATE WORK—Actual fire suppressant drops and rescue work can usually be performed by competent, current pilots.

Ground users of all night vision systems will benefit from knowledge of these principles. Until the user understands the interpretation of what is being observed, the information can be misleading. For example, a "hot-spot" detected by FLIR may turn out to be an object heated by solar radiation.

Operations

At project inception, a set of missions was developed to provide some direction to test activities. Not listed in priority, they were:

- Visual Reconnaissance
- IR Mapping
- Transportation to assemble and disperse overhead
- Burning out operations (firing ignition devices from helicopter)
- Emergency rescue missions
- Transportation of men, equipment and supplies
- Water and chemicals (fire retardant) delivery

It is interesting to note, that with the exception of transporting personnel and delivering ignition devices, all of the above missions have been operationally and experimentally carried out successfully and were truly legitimate needs for both day and night operations.

The initial experimental flying was performed by James Sanchez and Ted Hellmers of the LACFD. Also flying in the later stages of the program were Tony Hierholzer and Roy Cox of Western Helicopters, Inc., Rialto, Calif.

Current operational capabilities are maintained by the LACFD helicopter air unit. All pilots have been trained and are maintaining currency in night operations with systems provided by this project.

NVG Use

The first night viewing of an actual forest fire with NVG occurred on the Angeles National Forest, Calif., on February 12, 1974. This was the Devil's Canyon Fire which amounted to nearly 100 acres, and was diminishing during nighttime hours. The night vision crew flew into the area after 10:30 PM and practiced approaches into a small meadow adjacent to the fire. This was the first time that the crew had an opportunity to view actual burning material and hot coals through the NVG. They were very impressed with the distinction of the "hot spots" that were not visible to the naked eye. The entire periphery of the fire was visible with NVG. Men on the fire and their headlamps were easily seen; the supplemental lighting provided by the fire permitted easy visibility of the general area with the NVG. Later ground crew reports indicated that they could not see the helicopter in flight, although they could hear it fly over.

On June 16, 1974, flights were made on the Rock Fire in Texas Canyon on the Angeles National Forest and the first drops of water were performed for actual fire suppression. The basic technique of one pilot removing the goggles, and making the drop visually by the light from the fire, was developed on this fire. During the initial run on the fire, one pilot would remove the goggles, pick up the target,

Fig. 7. Soboba fire viewed through NVG.

and make the actual jettison of water. The second pilot would leave the goggles on, looking into the dark areas for the turnout after the drop. The control of the helicopter would then shift to him for flight away from the fire into the dark. The first pilot would then redon the goggles to assist during the flight back. Landings and takeoffs were made with both pilots wearing the goggles.

The most successful demonstration of night fire suppression capabilities occurred on the Soboba Fire, San Bernardino National Forest, Calif., on the night of August 28-29, 1974. After early evening reconnaissance flights, the Bell 204 helicopter was loaded with water in a 330-gallon "belly" tank at the fire camp heliport and started delivering water to targets at 9:30 PM. Operations continued until about 2:00 AM, with turnaround time averaging 4½ minutes. Total flight time for delivering water was 4-2/3 hours, with over 16,000 gallons delivered. All operations, including ground loading, were conducted under ambient light conditions provided by a 3/4 moon, or illumination from the fire itself.

Several things were learned from this operation:

- NVG can be used operationally for extended periods of time without exceeding normal fatigue limits.
- Light from the fire is an excellent supplement to ambient light, although the pilot has to avoid direct viewing of flames.
- Water at night appears to have superior suppressing and holding qualities over normal daytime use. This results from reduced winds and temperature, and increased humidity.
- Properly trained and experienced pilots are essential to safe operations.

During the calendar year 1975, the LACFD helicopter crews became fully operational with NVG. The following table shows that in over 40 hours of these operations, fire fighting constituted 44%, search and rescue 20%, training 25%, and demonstration flights 11%.

1975 Night Helicopter Operations, LACFD

Date	Hours	Description
1-18-75	1.0	Training—Bell test pilot
2-6-75	1.3	Recurrent training
2-9-75	.4	Night rescue—returned enroute
5-10-75	.6	Night rescue—Point Dume—victim transported
5-18-75	1.3	Night rescue in mountains—transported victim
6-13-75	1.3	Search for reported downed aircraft
7-11-75	.4	Simulated autorotations with "daytime" goggles
7-26-75	.8	Night recon of U.S. Forest Service fire
8-9-75	3.1	Pacoima Fire—24 water drops of 360 gallons each
8-18-75	1.4	Training C.D.F. pilot in 206B
8-19-75	1.9	Training C.D.F. pilot in 206B
8-20-75	1.9	Training C.D.F. pilot in 206B
9-13-75	1.2	Night rescue
9-22/23-75	9.9	Night fire Sierra Madre—flown VFR after recon
10-18-75	.5	Night rescue—Cienega Campground
10-27-75	.5	NVG demo flight for State Department
11-18-75	1.3	NVG demo flight for U.S. Forest Service, San Dimas
11-21-75	1.4	NVG demo flight for L.A. County Sheriff's Department

11-24/25-75	4.3	Water drops—night, Big T fire
12-9-75	.7	Recurrent training
12-10-75	.6	Recurrent training
12-15-75	1.5	NVG demo flight for U.S. Immigration Service
12-19-75	3.4	NVG search for missing U.S. Forest Service patrolman—found deceased in vehicle
TOTAL HOURS	40.7	

During the latter part of 1975, one pair of NVG was loaned to the Plumas National Forest for use in detecting nocturnal Christmas tree poachers. Although the formal report has not been received regarding the use, verbal reports indicate considerable success, and a definite deterrent effect resulted from the awareness of Federal Forest Officers using LLL devices. This factor may well be important in other field work until the public users outnumber the enforcers.

In the field of surveying, Forest Service experiments are now underway utilizing a transit-mounted LLL scope for direct viewing of vertical laser beams or "range-pole" at considerable distances. These laser range-poles are the surveyor's latest methods for accurate sighting from several miles away. Up to now, the receiver system utilized was all electronic, with sophisticated gating requirements to detect short low-power laser pulses. By using LLL scopes at dusk or night, this higher cost equipment is preempted. Also, lower power laser range-poles can be utilized, which further reduces equipment costs and size.

Now that Land Managers are aware of the capabilities of LLL equipment, there is a real need to conduct a screening program to spell out, in laymen's terms, the operating limitations of the equipment. Since there is now a proliferation of early design bulky systems competing with newer smaller higher performance devices, the buyer should be made aware of cost-benefits. Generally, once field users have tried the second-generation lightweight equipment, it is doubtful that they will be satisfied with anything else but further miniaturization.

Photography by use of LLL devices will be an important tool in the field. There is a need for LLL equipment designed specifically for this purpose.

Thermal Imaging System (FLIR) Use

Although the mini-FLIR did not become operational until the end of 1975, the Southern California fire season was still active and provided several opportunities for use of the FLIR system. Two major fires occurred on the Angeles National Forest in late November which were controlled with normal equipment and techniques. However, a week after the fires were declared "out," a helicopter flight was made around the periphery of the Tujunga Fire which disclosed several areas having hot material still glowing; and over the control lines! Crews were dispatched for digging out and extinguishing these areas. This same tactic was performed in January 1976, after a small 20-acre fire was declared out in the Clear Creek area of the Angeles National Forest. It appears that this may become routine for follow-up examination of recent burns to determine if hazardous firebrands still exist. In the past, this has usually been done at considerable cost by routine ground patrol.

Fig. 8. Laser range-pole LLL transit.

It is expected that there will be additional FLIR use during active fires to gather intelligence data through smoke and obscurations, as well as providing information for mapping and record purposes. The taping capability should also open up other uses for initial attack when the fire officer needs to have a rapid appraisal of the fire scenario.

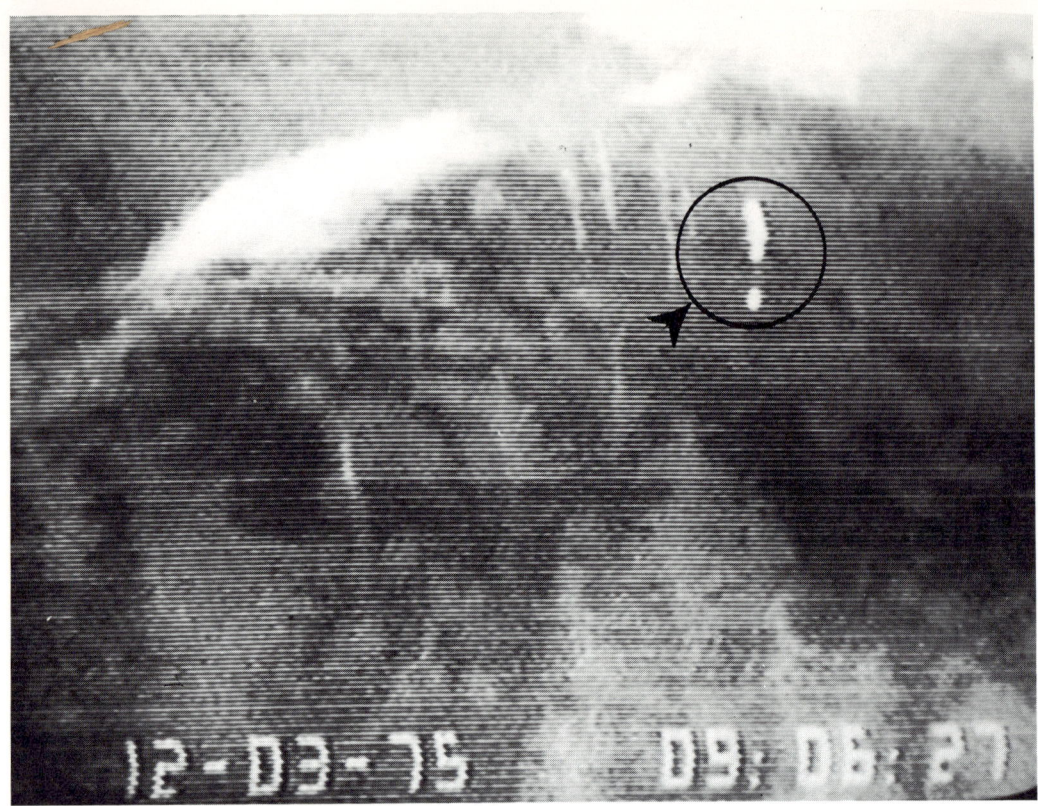

Fig. 9. FLIR imagery showing burning tree.

When additional FLIR systems are available, no doubt there will be experiments conducted for nocturnal wildlife studies, since considerable interest has been expressed by wildlife biologists for this use. This will require the addition of long-range viewing lenses to the sensor.

The use of night vision systems by the Land Manager has only begun. As soon as increased inventories reach the field, other uses are certain to be developed. Better management of land and natural resources has already resulted because of this technology.

Selected References

Ellis, W.E., "Helicopter Night Operations, Volume II, Effectiveness Analysis," Aerospace Corporation Report No. ATR-74 (7442)-1, Vol. 1, 1974.

Ellis, W.E., "Helicopter Night Operations, Volume III, Night Operations Implementation," Aerospace Corporation Report No. ATR-74 (7442)-1, Vol. III, 1974.

Rowe, J.A., "Helicopter Night Operations, Volume I, Night Vision Devices," Aerospace Corporation Report No. ATR-74 (7442)-1, Vol. 1, 1974.

Shields, H.J., "Night Helicopter Operations, Information for Flight Crews," USDA Forest Service, August 1975.

UNUSUAL APPLICATIONS OF IMAGE INTENSIFICATION DEVICES

Thomas M. Brennan and William H. Dyer
Baird-Atomic, Inc.
125 Middlesex Turnpike, Bedford, Massachusetts 01730

Abstract

Baird-Atomic has developed and marketed image intensification systems to the U.S. Military and commercial customers. In order to be successful in the latter market, maximum utilization must be made of military-funded components, existing sales and service organizations, and company commitment of personnel and front-end money. Several examples of commercial projects are examined and contrasted to military night vision programs.

Introduction

Baird-Atomic has been a systems design user of image intensification tubes for over 10 years. Like most companies in this field, we got our start through military sponsorship. In addition to the image intensification tube, most of our commercial projects have included optical subassemblies developed for the military. In this paper, we shall discuss the technical and marketing aspects of several of Baird-Atomic's commercial night vision projects.

Background

In the mid-1960's, Baird-Atomic was one of the first contractors to outfit an airborne viewfinder with the first generation, 25-millimeter, three-stage, electronically focused image intensifiers. Although there were delays in starting, the Navy had a detailed specification and knew pretty much what the performance would be before the project was started. As you will see from Fig. 1, the RA-5C Day/Night Viewfinder was a very sophisticated electro-optical device—complicated primarily by the physical constraints of bringing the image from beneath the aircraft up to the navigator's eye. The system was very costly in development and test, most of which was paid for by the Navy. We will contrast this to a commercial project for a similar device.

Night Fish-Finder

During the day, commercial fisheries utilize spotter aircraft to search and find schools of fish. A commercial fishing organization came to Baird-Atomic to develop a night viewfinder to detect schools of fish through the phenomenon of bioluminescence (light emitted by the mechanical agitation of the dinoflagellates plankton by the motion of fish). This light emission is a form of chemiluminescence with a typical spectrum shown in Fig. 2. This light is visible to the naked eye at night.

The basic design considerations of the development were low cost, sensitivity, and directable line of sight in azimuth and elevation. The device was fitted into a Cessna 337 pusher/puller, twin engine aircraft equipped with an 18-inch-diameter camera port on the bottom of the fuselage, as shown in Fig. 3. Utilizing the existing camera port minimized the problems of getting the aircraft certified for flight by the FAA.

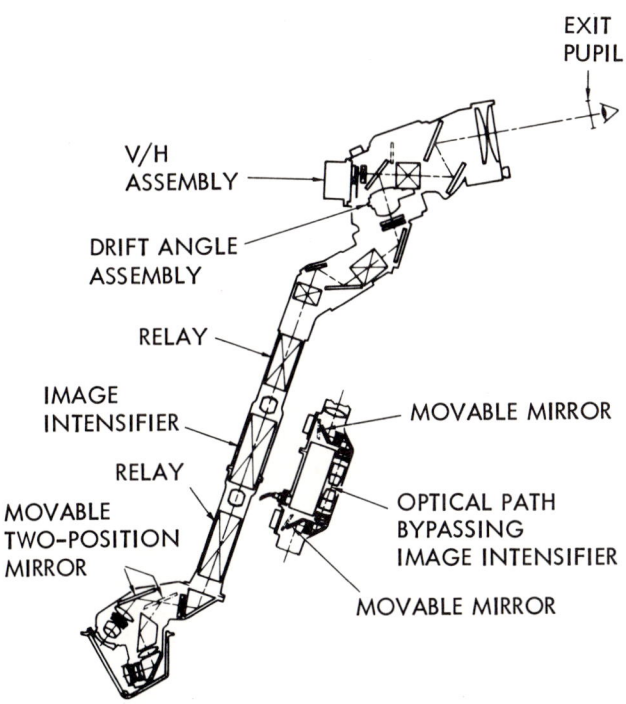

Fig. 1. RA-5C Day/Night Viewfinder schematic.

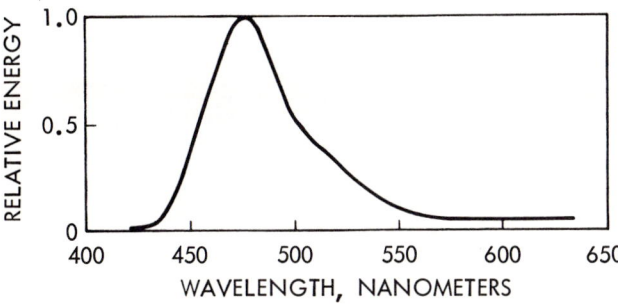

Fig. 2. Bioluminescence spectrum.

The device consisted of a vibration isolating interface bolted to the floor of the aircraft, from which were suspended two continuous azimuth rotation bearings to support the viewfinder (Fig. 4). The optical components are depicted in Fig. 5. The two flat plate windows were mounted into the turret, supported on three edges by the turret structure and joined by a silicon adhesive without supporting structure at their interface to reduce the vignetting of the collecting optics. The large, double dove prism (see Fig. 5) used for elevation scan was truncated on the corners to reduce the weight and size of the turret. Two sections of the prism assembly were polished and aligned to 0.25 minute of arc. The prism directed the line of sight to an f/1.2, 160-millimeter-focal-length catadioptric lens. This lens was originally developed by Baird-Atomic for the U.S. Army AN/TVS-5 Crew Served Weapon Sight program. Although designed for the 25-millimeter-diameter format, the lens provides usable imagery out to 35 millimeters. For this reason and for the additional length and field of view, a 40-millimeter first generation image intensifier was selected. The image tube used was selected for its response in the 470-nanometer wavelength range and for a luminance gain of 100,000 when tested with a 2870° Kelvin source. The green P-20 phosphor output of the image tube was viewed through a biocular eyepiece developed by Baird-Atomic for the Night Vision Laboratory AN/TVS-4 NOD. This eyepiece magnified the format of the image tube 3.7 times and permitted dual-eyed viewing of the entire phosphor, necessary for the in excess of 4-hour flight missions intended. The system specifications achieved were:

Resolution	35 lp/mm
Magnification	2×
Field of view	12 degrees
Electrical power	6.7-volt battery
Coverage	180-degree hemisphere beneath the aircraft

The only operator controls were the on/off switch mounted on the turret, the elevation handles for direct drive of the prism, and total azimuth rotation.

During the flight tests, some problems became evident. The engine exhaust dirtied the windows and caused increased background brightness at high elevation settings when looking forward. This was solved by rerouting the exhaust around the turret. At the fishing site, some problems developed which we were not made aware of initially. In the fishing area, there were offshore oil wells which have lights and burn excess gas, increasing the background and reducing the contrast between the bioluminescence and the sea. This produced ghost images within the double dove prism and unwanted false signals. The ghosts were exorcised by increasing the baffling external to the double dove prism and by blocking the path between the two sections of the prism. The unwanted energy from the lights and fires were eliminated by an unusual process. Narrow bandpass filtering was rejected because it would have reduced the weak signal in the bioluminescence spectral range. The solution was to replace the broad band silver coating on the primary mirror of the catadioptric lens with a slightly detuned laser reflectance coating peaked at 470 nanometers. The resulting system spectral response can be seen in Fig. 6. Curve A shows that the system was originally dominated by the S-20 photocathode spectral response. Curve B is the system response with the selective mirror. Curve C is the short wavelength end of the 2000° Kelvin blackbody source. There is an 8 to 12 times improvement in the rejection of the near infrared and visible background caused by the light and gas flame sources. In addition, the absolute reflection in the desired signal wavelength region is improved. Once these improvements were implemented, the system worked very well.

Fig. 3. Cessna 337 with viewfinder turret.

Fig. 4. Fish-finder mounted in camera port.

The marketing lessons learned were the customer's great technical and monetary interest in this equipment which he planned to use to achieve financial gains for his company. Although the customer did agree to pay for some engineering design costs, he was unwilling to increase the price to solve problems, such as the lights and gas flames. The ultimate performance of the system was much more important than his initial specifications. It should be noted that although we knew this was a one-shot deal, we had hoped to sell a similar system to the Air Force to be given to the South Vietnamese. However, this market went away.

Security Market

Baird-Atomic has participated in several other of the commercial markets involving image intensification. In the police and security markets, Baird-Atomic has endeavored to market a direct-view image intensification device. However, the main competition is always the commercially available objective optics which the market is not sophisticated enough to reject. Distribution, service, and warranty are also very important factors in this market which are not required in military business. Utilizing the sales force of an organization already selling bullets, badges, and billy clubs is very desirable.

Baird-Atomic was one of the first firms in the x-ray baggage inspection business (Fig. 7). Our initial systems were sold to the Customs Bureau to a detailed specification under a competitive development contract. However, the airport baggage inspection systems did not follow this scenario. The FAA did not come out with detailed specifications for baggage inspection systems, nor did they buy the equipment. The market was to follow very commercial lines. The firms that were in this market developed the products and sponsored demonstrations for various commercial airlines. The aspects of service and warranty were very important, and, in the beginning, cost was much more important than the radiation safety of the equipment. Each of the four systems sold by Baird-Atomic to Government specifications were in excess of $60,000 apiece. The current price for airport inspection systems is about $20,000.

Underwater Viewer

Figure 8 shows the Baird-Atomic AN/VVS-2 Night Driver's Viewer for Combat Vehicles. Although we do not expect a commercial market for this device, with the possible exceptions of Boston and Rome drivers, we have had success with using this device in an underwater application. One of our vendors is also an underwater salvage outfit. They wanted to determine whether this type of device would be useful in underwater search and inspection applications. The concept being that it would be used during a broad inspection and then lights and camera equipment or whatever would be brought down when specific points of interest were found. Although we feel there is a market for this type of equipment to underwater salvage outfits, marine biologists, etc., we feel there is a considerable amount of development costs in coming up with a product, and it would also require marketing and service through existing channels servicing these markets.

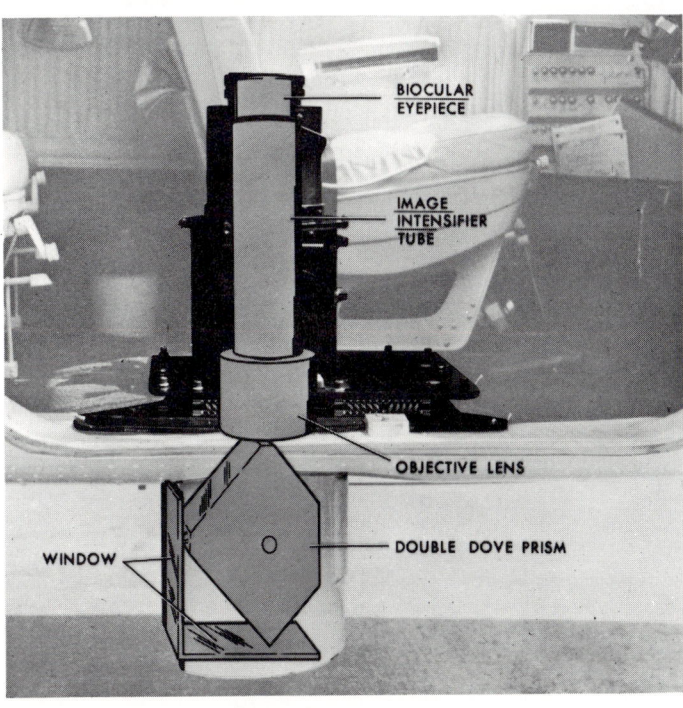

Fig. 5. Fish-finder optical schematic.

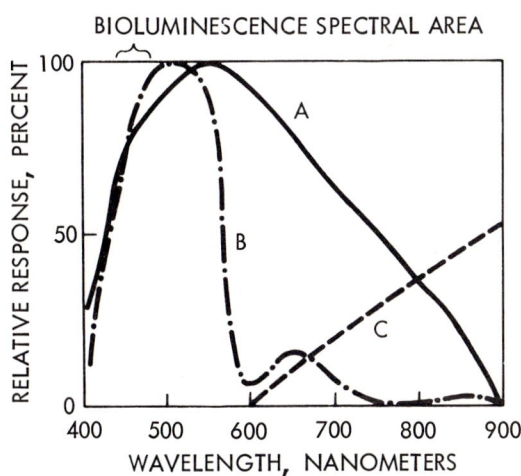

A. ORIGINAL SYSTEM

B. SYSTEM WITH NARROW BAND REFLECTIVE MIRROR

C. SHORT WAVELENGTH OF 2,000 °K BLACKBODY CURVE

Fig. 6. Fish-finder system response.

Selected Wavelength Imaging

Baird-Atomic has developed several direct-view image intensification concepts using selected wavelength imaging. For example, a narrow bandpass filter could be incorporated into the image intensifier to look for specific pollutants or compounds in air or water emissions. However, in addition to all the normal marketing problems, one would also add the legal validity of such a device. In our in-house operation of assembling optical emission spectrometers, we are required to precisely align for elements which emit in the near infrared. Although we do not envision selling this type of equipment which we are using in our commercial final alignment and assembly, there may be other in-house applications for which image intensifiers can be used.

Conclusions

These and other competition sensitive commercial applications of image intensification are not limited to law enforcement and medicine, but by man's imagination. We in industry who deal with the military are used to dealing with technically qualified personnel who are prepared to commit front-end money to develop state of the art devices. Schedule delays and cost growths can be tolerated. Commercial customers tend to be nontechnical, business people who want to see the hardware first, have little or no nonrecurring funds, and schedule delay may mean financial loss as well as personal embarrassment to the individuals. Money can be made in a commercial marketplace, but you must take maximum advantage of military developed components and face up to the fiscal and personnel requirements of product design, marketing, and service.

Fig. 7. Fail-Safe III x-ray baggage inspection system.

Fig. 8. AN/VVS-2 Night Driver's Viewer.

Session 3
APPLICATIONS OF LOW LIGHT LEVEL DEVICES IN ASTRONOMY

Session Chairman
W. Kent Ford
Carnegie Institution of Washington

A LARGE IMAGING ARRAY CCD PROGRAM

Fred E. Vescelus
Jet Propulsion Laboratory
Pasadena, California

Gault A. Antcliffe
Texas Instruments
Dallas, Texas

Abstract

Much effort has recently been devoted to Charge Coupled Devices (CCD's) as imaging detectors. This paper describes a large imaging array CCD program currently underway at the Jet Propulsion Laboratory and Texas Instruments and presents test results obtained on both 100 x 160 element and 400 x 400 element arrays. Expected low light level performance is also given, along with a brief description of future plans.

Introduction

Several years ago, charge coupled devices (CCD's) revealed themselves as potentially superior visual imaging sensors. At that time, a program was conceived to evaluate the various approached that could (at that time) be considered; select the most promising one; and proceed, in a phased development, to derive large imaging array sensors.

A partial list of the various technology approaches considered is contained in Table 1, and includes such things as front vs. back-side illumination, surface vs. buried channel, and the transfer electrode configuration. The potential performance capability, the current state of development, the implementation ease, and the radiation susceptibility were all considered in making the final selection.

Table 1. CCD Technology Approaches

Area		Pros	Cons	Selected
1. Front vs. Backside Illumination	Front	• Simple • Existing Technology	• Lower Sensitivity • Interference Filter Effect	
	Back	• High Sensitivity • No Interference From Polysilicon • Potential UV Response	• Complex Thinning • Less Rugged	X
2. Gate Material	Polysilicon	• Transparent	• Poor Radiation Hardness	
	Aluminum	• Lower Temperature Processing	• Opaque	X
3. Surface or Buried Channel	Surface	• Simple • Bloom Control • Low Dark Current	• Poor CTE • Noisy • Radiation Sensitivity	
	Buried	• High CTE • Lower Noise • Less Radiation Sensitive		X
4. Number of Phases	2φ	• Fewer Transfers • Smaller Pixels	• Lower Dynamic Range • Uni-Directional	
	3φ	• Greater Dynamic Range • Bi-Directional	• More Transfers • Larger Pixels	X

After due deliberation, it was decided to select a thinned back-side illuminated, buried-channel, three-phase CCD approach. The thinned, back-side illuminated approach eliminates imaging through a polysilicon gate structure which, in turn, eliminates the optical interference filter effect created by the multiple layer gate structure. The image is projected directly on the backside of the silicon, avoiding this problem.

The buried channel array has lower inherent noise due to elimination of the surface trapping states, and better expected radiation resistance. The three-phase approach provides for a straightforward clocking scheme that can shift data in either direction, a larger full-well capacity with fewer diffusions or implantations than a two-phase device.

Effort was first channeled into verifying that the back-side illuminated, buried-channel approach held promise for large array development by initially fabricating 100 x 160-element arrays as an approach verification and process refinement and vehicle. Figure 1 is an example of the output quality obtained.

The 100 x 160 arrays were very successful and demonstrated that, although a novel approach was selected, the techniques required to implement this approach were understood and practical, and the results justified the backside illuminated approach. Also, at this time, considerable effort was being directed into studying the noise performance of the CCD sensor and associated signal processing electronics. At the 100 x 160 array level, these approaches included a balanced sample and hold configuration and a correlated clamp amplifier.

Work then proceeded on a 400 x 400 device, using the experience and approaches developed from the 100 x 160 program. This device also uses buried-channel, back-side illumination, and three-phase configuration. The readout approach selected is a simple precharge amplifier, chosen as a simple output circuit to optimize device yield.

Current Performance

The 400 x 400 array device is in existence and performing well. While there is still work to be done, this device has already proved to be an excellent performer. The device is seen in Figure 2.

Fig. 1. Imagery from 160 x 100 buried channel ccd.

Fig. 2. 400 x 400 ccd array.

First and foremost is the overall imaging performance of the array. Figure 3 depicts the imaging performance. As can be seen, the overall output is very good, with one dark current streak visible at a frame rate of 163 milliseconds, and one blocked channel.

The spectral response measurements made to date have been restricted to between 0.45 and 1 μm because of experimental limitations. The results obtained, however, look very promising and show a quantum efficiency of greater than 70% at 0.7 μm wavelength. The spectral response curves obtained to date are shown in Figure 4. Additional work is underway to extend measurements below 4000 Å.

The dark current generation rate observed in the 400 x 400 devices has been measured over a 70°C range and found to be very low.[1] With cooling, the device can be used to integrate and/or read out over

---

1. For example, JPL #11, (a 400 x 400 device) has 0.9 nA/Cm^2 dark current at 240°C.

Fig. 3. Imagery from 400 x 400 ccd.

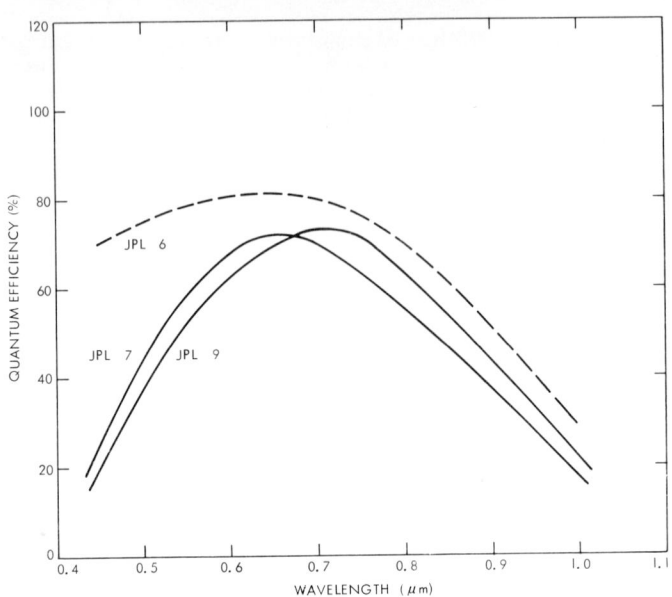

Fig. 4. CCD spectral response.

periods of seconds to hours, depending upon the cooling and the allowable dark charge. The dark current generation rate observed is given in Figure 5.

The classical technique to measure linearity is to measure the device response vs. exposure. This technique was tried, but the precision in our initial data did not allow the conclusion that the devices were, in fact, linear on a wide range of input power. Therefore, a technique was developed to measure a small fixed change in response as a function of the total exposure. This technique was used, and yielded a linearity coefficient of 0.9987 over the device's dynamic range. The response nonuniformity was measured over a 400 pixel area on JPL #10 sensor, and found to be ±2.7%, with a dark current nonuniformity at -40°C less than the 100 electron ims noise.

The device resolution has been measured using the edge response generated by a moving knife edge. The horizontal sine wave response measured by this technique is given in Figure 6, along with the

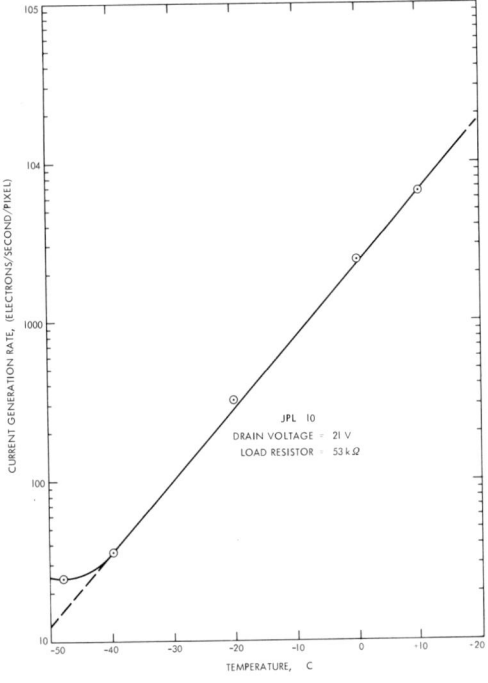

Fig. 5. Dark current generation rate.

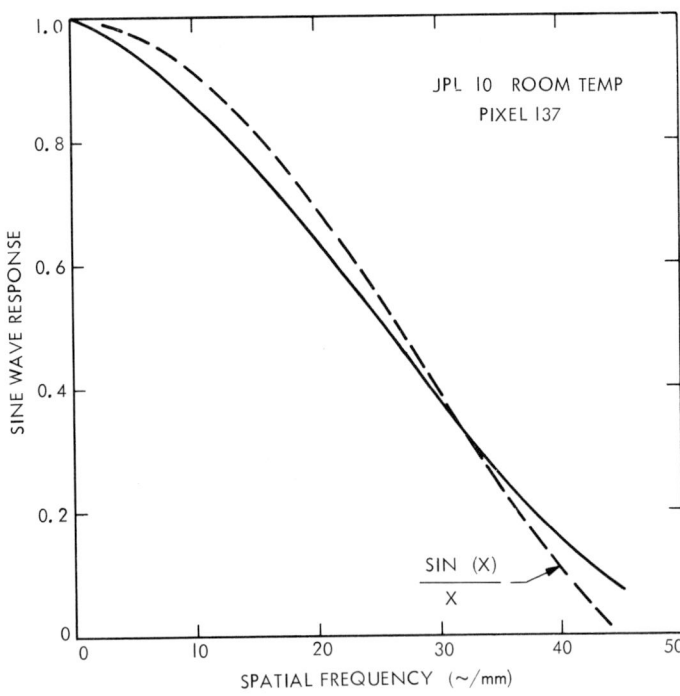

Fig. 6. Horizontal sine wave response.

theoretical sin(x)/x curve, and as can be seen, the measured performance closely approximates the ideal curve. Figure 7 shows the horizontal line-spread function obtained with the same device. The transfer efficiency of these devices has been measured and is about 0.99995.

Much work has gone into understanding and minimizing the readout noise associated with the CCD, since this is one of the major limitations on the ability to perform low light level imaging. Both on-chip and off-chip electronics have been used together to give pixel readout noise less than 50 electrons for the 400 x 400 device. Readout noise substantially less than this level is anticipated with further array development.

Low Light Level CCD Performance

The performance obtained to date with the CCD's tested has been used to determine the performance expected under lower light conditions. Using data obtained at Texas Instruments along with the low noise readout (50 electrons) at JPL, the curve of Figure 8 was generated. This curve compares a 250 x 500 element direct view (non-intensified) CCD with both the 25mm and 80mm intensified SEC vidicon. While, at very low light levels, the CCD does not perform as well, the large dynamic range of the CCD can be used in many applications to great advantage.

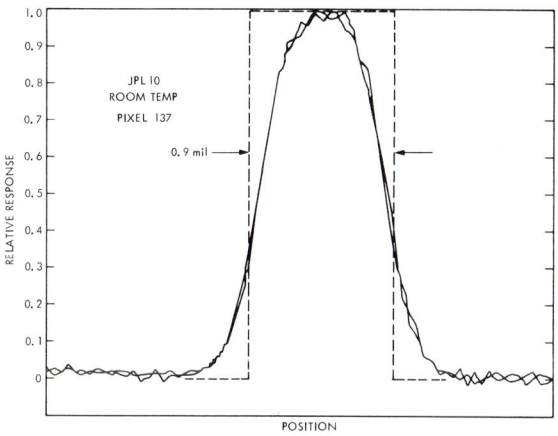

Fig. 7. Horizontal line spread function.

Fig. 8. Video signal to noise ratio.

The low light level performance can also be displayed as a function of limiting observer resolution. This data, based on the signal to noise curves of Figure 8 is shown in Figure 9 which includes, in addition to the two 1-SEC vidicons, CCD performance for no readout noise and 30-electron readout noise, and silicon vidicon performance. These curves assumed a pixel of 0.9 + 1.3 mil for the CCD. Two important points need to be made here. One is that, unlike the photon-noise-limited sensors (the intensified devices), the CCD is readout or electronics noise limited at low light levels and, therefore, by increasing the collection area on the chip (the pixel size), the photon collection can be increased without additional readout noise. This can be used to translate the CCD performance curves to the left, as a direct function of the pixel size. The second point is that unity MTF has been assumed for these curves. In reality, the MTF rolls off at higher resolutions for all these devices.

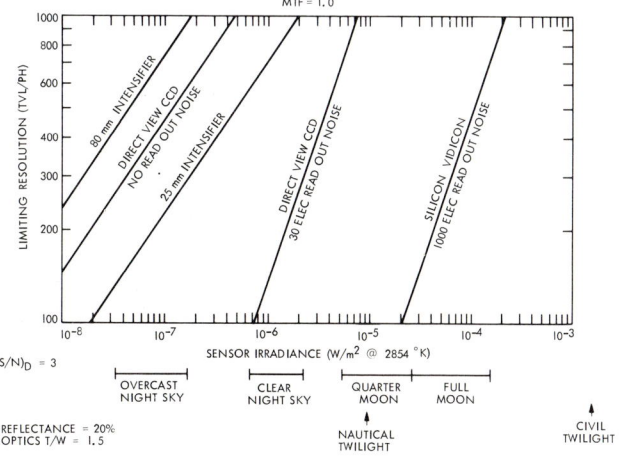

Fig. 9. Noise limited resolution

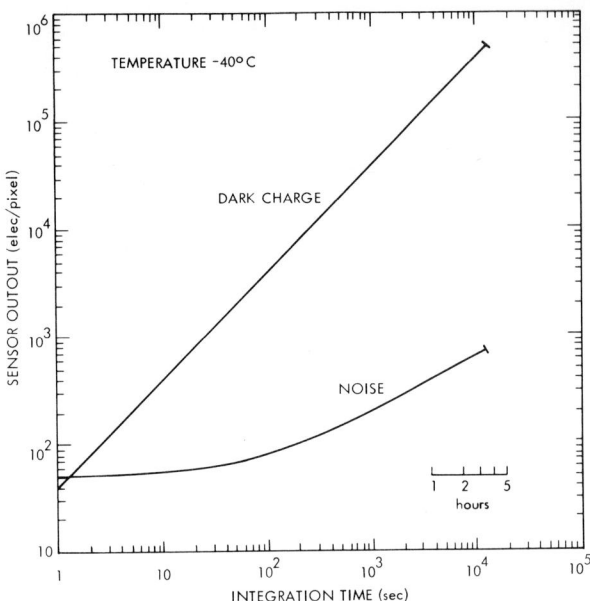
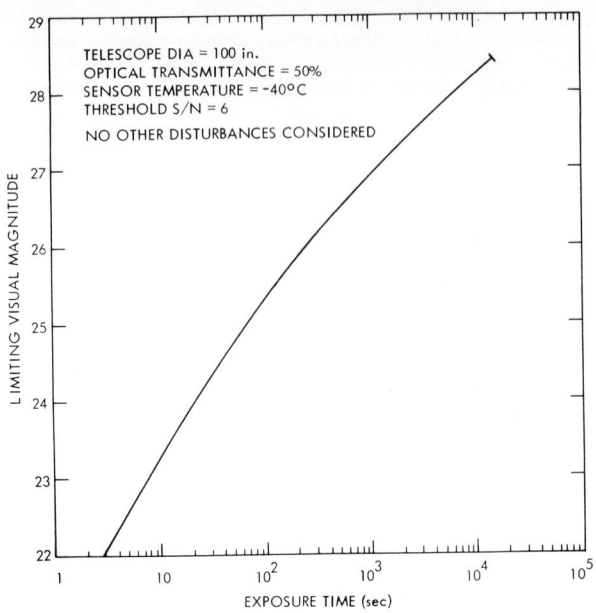

Fig. 10. Dark Output and Noise for 400 x 400 element ccd.

Fig. 11. Star Detection Capability of 400 x 400 element ccd.

The low light performance can also be displayed as a function of star detection capability. To do this, the dark current data in Figure 5 is plotted, along with the combined sensor dark current and readout noise, in Figure 10. The full-well capability of the current CCD's is around 5×10^5 electrons, so times up to device saturation have been plotted. The device, at -40°C, takes approximately 5 hours to saturate. The noise considered here is the random noise; to take out the element-to-element variations, which become more pronounced at long storage times, requires data decalibration.

The noise and dark current data in Figure 9 can now be used, along with stellar photometric data,[1] to determine the possible performance of the CCD for star detection. Performance, in terms of star detection capability, is plotted in Figure 11, and does not include any observational limitations imposed by such things as the night sky and tracking variances. As can be seen, the CCD with its long storage time capability, can detect faint objects and should be a valuable tool in future astronomical work.

Future Plans

The 400 x 400 CCD sensor will continue to be refined. In addition, work is underway to improve the UV response, reduce the readout noise, and increase the array pixels to the 800 x 800 region.

References

1. G. A. Antcliffe, L. J. Hornbeck, W. C. Rhines, W. W. Chan, J. W. Walker, and D. R. Collins, 1975 International Conference on the Application of CCD's, San Diego, California.

2. G. A. Antcliffe, L. J. Hornbeck, J. M. Younse, J. B. Barton, D. R. Collins, JPL Symposium on CCD Technology for Scientific Applications, March, 1975, Pasadena, California.

Acknowledgment

This paper presents the results of one phase of research carried out at the Jet Propulsion Laboratory, California Institute of Technology, under contract NAS 7-100, sponsored by the National Aeronautics and Space Administration.

---

1. Stellar photometric data for various photocathode materials, Forbes and Mitchell, Lunar & Planetary Laboratory, U. of Ariz., Oct. 25, 1968.

ASTRONOMICAL APPLICATIONS
OF CHARGE INJECTION DEVICES

R.S. Aikens, C.R. Lynds and R.E. Nelson
Kitt Peak National Observatory
950 North Cherry Avenue
Post Office Box 26732
Tucson, Arizona 85726

Abstract

During the past year several types of silicon self-scanning arrays have become available. The Charge Injection Device (CID) which is processed in a 100 x 100 format has been examined in detail and is presently being used on a regular basis at Kitt Peak National Observatory for a special class of astronomical problems. The random address and non destructive readout features make the CID a unique member in the family of array devices. Linearity, MTF, and Quantum Efficiency have been measured and detailed results are given.

Introduction

The General Electric Company introduced the Charge Injection Device in 1974. Since that time CCD's have dominated the solid state array scene, while the CID has received little attention. The Kitt Peak National Observatory has been testing several solid state imagers for use on astronomical problems, and the CID exhibits several properties which render it unique. In general a photo detector must yield reproducible, high quality output if it is to be a useful photometric aid to the astronomer. By this definition the CID is well-suited for a variety of photometric problems over a wide range of operating flux levels.

I. Device and System Description

The Charge Injection Device is conceptually simple: an array of pixel sites, each consisting of a dual MOS capacitor. In the array configuration, one capacitor plate is associated with a row and the other with a column. Through prescribed application of voltages and the use of appropriate sensing circuitry, a site can be made to integrate carriers generated by photon excitation, be read out, and re-established for a new integration. Several CID site states are illustrated in Figure 1. At a site a negative

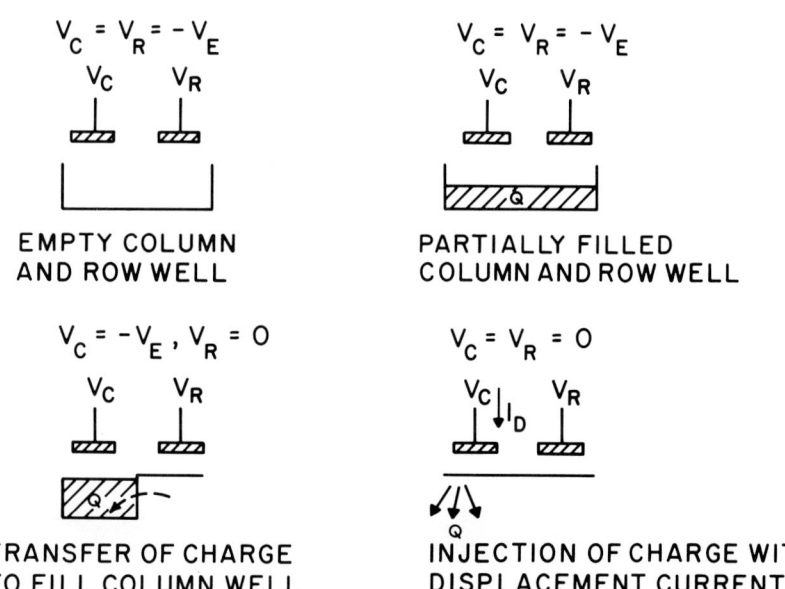

Figure 1. Pixel site states using potential well concept.

potential can be applied to either or both capacitor plates, resulting in the storing, transferring, or injection of the charge. The charge stored at the surface consists of minority carrier in N type silicon (holes) and the surface is always biased sufficiently for inversion in order to avoid oxide-silicon interface state interactions.

The readout process starts with all rows biased to -15 volts and the columns at -7 volts. After the selection of a new column, a switch momentarily clamps the column line to -7 volts and allows it to float. At this time a 15 µ-second integration is made on this level. The selected row is then switched from -15 volts to 0 volts, causing all stored charge which existed in the row well to be transferred to the column well. Since the column line is floating, this charge will alter slightly the floating well potential. A second integration is made on the signal, this time inverted, so that the final value of the integrator is the time integral of the voltage increment due to the charge transfer.

At this point the charge is still under the column capacitor and can be either injected by reducing the column voltage to zero, or returned to the row well by dropping the row potential back to -15 volts; the latter method results in a non destructive readout (NDRO). This method of readout appears to provide a very powerful approach for increasing signal to random noise levels through repeated reads which are summed to produce the final data.

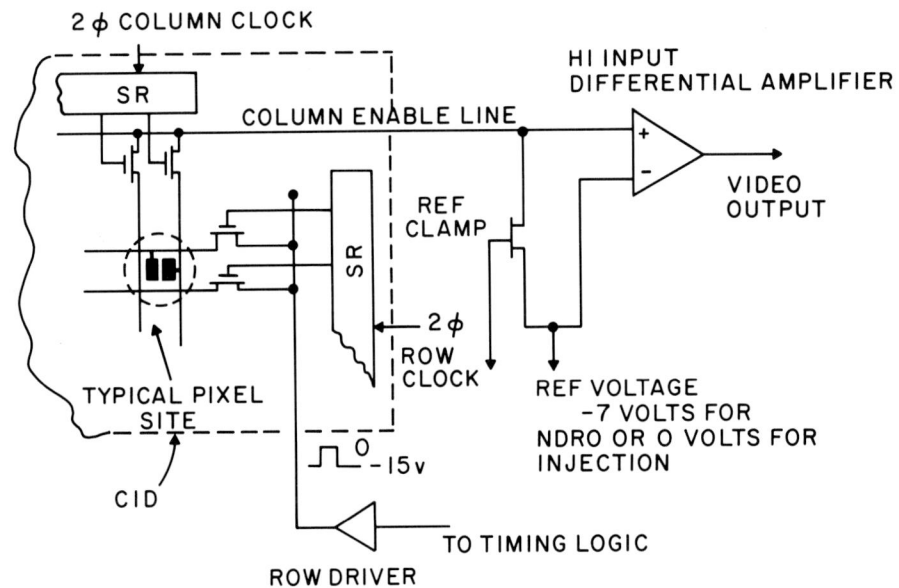

Fig. 2. CID readout technique for NDRO or injection.

Figure 2 shows the simplified circuitry used to operate in either mode and Figure 3 is a timing diagram for a single pixel. The column shift register is operated to give a pixel to pixel time of 50 µ-seconds, allowing time for the data to be encoded to 14 bits and read into a memory buffer. The row shift register is operated at rate 1/100th of the column clock rate. A readout takes 0.4 seconds, allowing time at the end for some computer setup for the next frame.

Several noise sources deserve comment, the most significant being the first stage amplifier. No attempts to use a low noise front end have yet been made; rather emphasis has been on development of readout methods and a reliable operating system. The maximum signal corresponds to about 6.0×10^6 charges, which produces an incremental output voltage across the column capacitance C_c, typically 25 pf, according to the relation:

$$\Delta V = \frac{\Delta Q}{C_c} \tag{1}$$

Substituting for C_c and full scale charge yields:

$$\Delta V \text{ max} = 40 \text{ mV}. \tag{2}$$

Another noise source is that associated with charge transfer between wells. The amount of charge moved will depend on applied voltages which have an uncertainty associated with their value. KT/C noise introduced by switches on the chip introduce another uncertainty. At this time it is felt that all noise sources, including the preamp, combine to produce a total system noise of less than 1000 electrons, most of which originates in the preamplifier.

While the selected row - column capacitance is a function of applied voltage, the

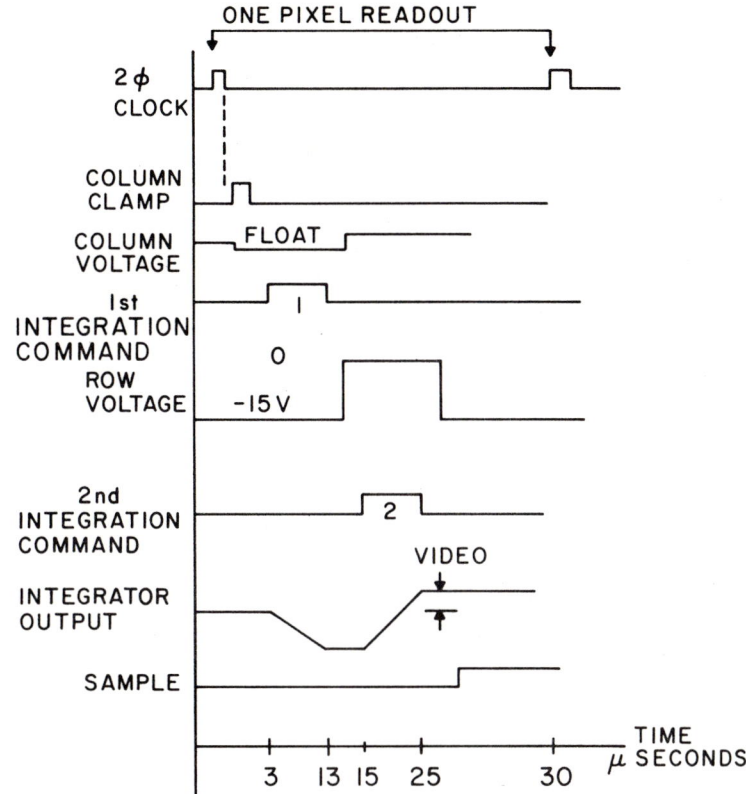

Figure 3. CID pixel readout timing.

action of all the empty column capacitors in parallel and the small voltage increment ΔV makes the final error negligible.

Figure 4 shows the basic CID system centered around a Varian 620 F Computer. Interactive "on line" picture processing is now routinely done while new integrations are being made. It is possible to periodically check an integration non destructively to see if a desired signal level has been attained. The readout mode then provides as many

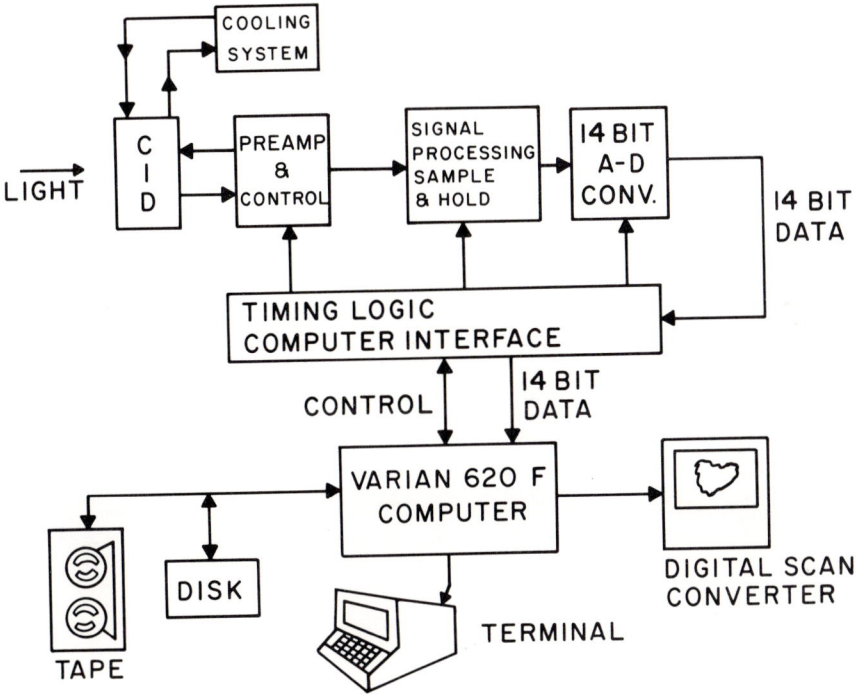

Figure 4. CID operating system.

non destructive reads to be made as deemed necessary to produce the required signal to noise ratio. In addition flat field gain calibration frames and dark frames are stored for final picture processing.

Figure 5. Example of signal to noise enhancement through repeated NDRO's.
Upper Left: High illumination reference frame.
Upper Right: Single frame of a low level illumination .
Bottom: Summation of 100 low level single frames.

Figure 5 illustrates the value of the NDRO process as applied to a very low light level exposure. For this test a bias is first exposed and non destructively summed 100 times. A scene is introduced to produce a 1:1 signal to noise ratio in the video. The CID is again read out 100 times and the data are summed. Finally the bias signal is subtracted from the summed data to produce a picture with signal to noise improvement of $\sqrt{100}$. Since the system noise is around 1000 electrons, it is believed that the noise in the final picture is improved by about $\sqrt{100}/\sqrt{2}$; $\sqrt{2}$ due to the subtraction process which sets the noise at about 140 electrons.

Experiments indicate that for effective NDRO operation, a small bias charge must reside in the potential wells. This charge must be present to fill surface states and amounts to about 3×10^5 carriers, with a corresponding shot noise of 550 carriers. The bias charge can be introduced optically with a flash of light of prescribed duration and intensity, or the charge can be injected to a specified residual which serves as a bias. Failure to maintain a bias results in some charge which is trapped and cannot be manipulated to provide non destructive readout.

The operating mode presently used requires the introduction of a flash before the exposure. It appears that the bias simply allows charge above a certain threshold to be mobile, and charge below that threshold to be trapped and immobile. This surface inversion effect is discussed in detail by Richman[1.].

A procedure for subtracting out the bias charge shot noise (550 carriers) has been devised as follows: Prior to an exposure the device is exposed to a flat light source which puts in a bias charge. Several hundred NDRO frames are summed and stored, and the scene illumination is introduced. After the scene has been exposed to the desired level, it is also readout non destructively and summed. Next the bias sum and shot noise component is subtracted from the scene, leaving only scene data.

II. Measured Properties

The fundamental CID properties which we have measured are:
1. Quantum Efficiency vs. Wavelength
2. Resolution
3. Linearity
4. Dynamic Range
5. Non Destructive Read Efficiency

---

1. Richman, MOS Field Effect Transistors and Integrated Circuits.

6. Dark Current
7. Row Column Crosstalk

1. Quantum Efficiency

Quantum efficiency measurements were made using the McMath Solar Telescope and high dispersion spectrograph. This is an excellent monochromatic source, and is well calibrated when used with integrated sunlight. Figure 6A shows the quantum efficiency versus wavelength over the range of sensitivity for the bulk and epitaxial arrays.

When exposed to a monochromatic source (100 mÅ/mm), interference fringes were observed due to the faceplate in the epitaxial device and the silicon substrate in the bulk device. The fringing increases red-ward to 1.1 microns where it modulates the signal by 50%. Figure 6A gives the results of quantum efficiency tests.

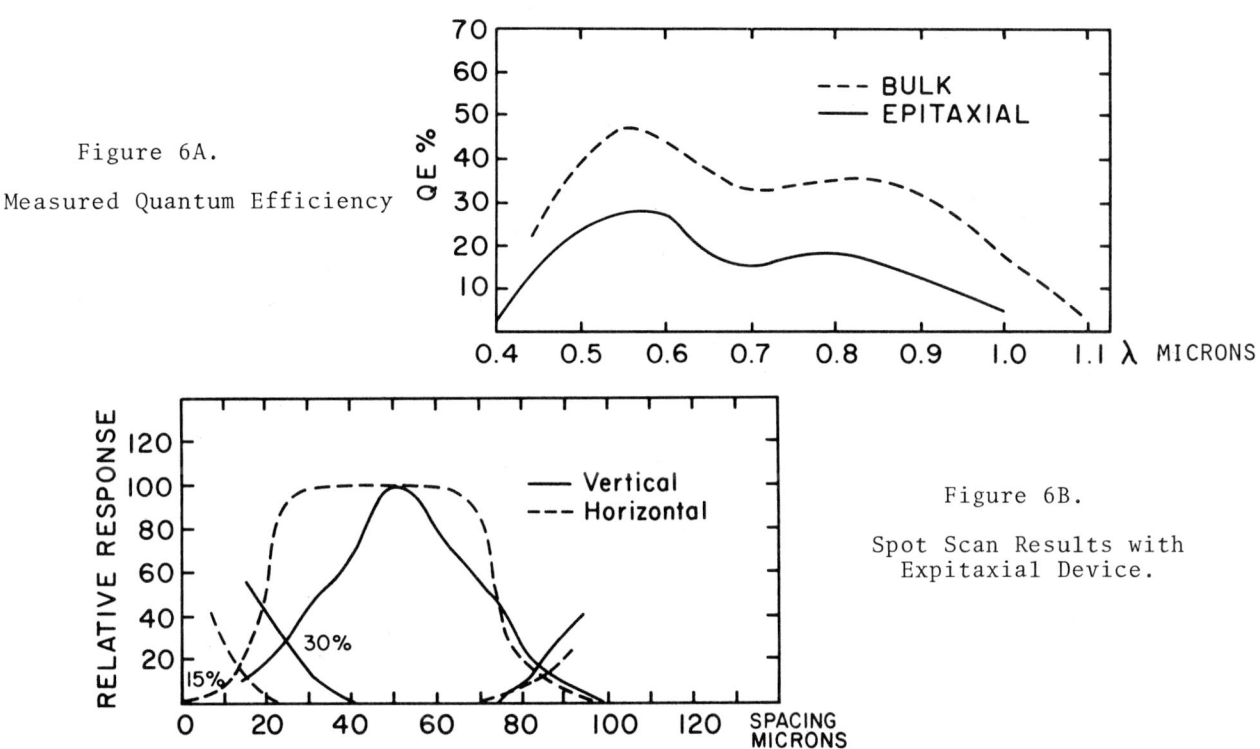

Figure 6A. Measured Quantum Efficiency

Figure 6B. Spot Scan Results with Expitaxial Device.

Figure 6. Quantum efficiency and resolution data obtained under laboratory conditions.

2. Resolution

Resolution was measured on a spot microdensitometer which produced a 2-micron spot. Both the epitaxial and bulk devices show excellent MTF when operated at 77°K. The MTF on the bulk CID is degraded by about a factor of 2 at room temperature. Figure 6B shows results obtained on the epitaxial CID and is also applicable to the bulk CID at 77°K.

3. Linearity

System linearity was measured by making 1-frame integrations on a source which was turned on for a prescribed time at a constant intensity. The non destructive readout allowed the signal to be measured after the exposure to each flash. A plot of first differences versus flash number was made and the CID amplifier combination was linear to ±0.25% with the non-linearity appearing as a slow smooth curvature, easily calibrated.

4. Dynamic Range

Full scale saturation to RMS frame-to-frame noise on a given pixel was measured at 6000 : 1. The present full scale signal is 6.0×10^6 charges and RMS noise is 1000 charges. The present input amplifier has noise level of around 20 nV/$\sqrt{HZ}$, which can be improved a factor of 4 using standard techniques to 5nV/$\sqrt{HZ}$. It is probable that other noise sources will come into play before 5 nV/$\sqrt{HZ}$ can be realized. $KT/_C$ noise from the column clamp is reduced to a small level by the double integration differencing technique (double correlated sampling.)

5. Non Destructive Read Efficiency

A high screen illumination test pattern was read non destructively for 9000 frames at 77°K. The first and 9000th frames were subtracted and the residual was measured. No structure was present and the difference appeared to be white noise at the preamplifier level. This translates into a loss of not more than 10^{-3} carrier/pixel/readout and possibly much fewer. It is impossible to see scene degradation while observing the monitor.

6. **Dark Current**

Dark current was measured at room temperature by observing the time required to fill a well in the absence of light. Typically this varied between 2.8 and 4 seconds.

Dark current (I_D) is the ratio of saturated charge to time required to reach saturation:

$$I_D = \frac{3 \times 10^6 \times 1.6 \times 10^{-19}}{4} \text{ coulombs/seconds} \tag{3}$$

$$I_D = 1.2 \times 10^{-13} \text{ amps} \tag{4}$$

The room temperature dark current density (I_p) is the ratio of dark current in a pixel to pixel area, and is:

$$I_p = I_D/A_p \text{ (where } A_p \text{ is a pixel area } 4.8 \times 10^{-5} \text{cm}^{-2}), \tag{5}$$

which gives:

$$I_p = 2.5 \times 10^{-9} \text{ amps/cm}^{-2}. \tag{6}$$

In normal use the device is cooled to 77°K and dark current is not significant except in an apparent hot spot on the right side of the array, which writes up slightly in a few hours. Cooling with liquid nitrogen totally eliminates dark current problems for integrations up to one hour.

7. **Row Column Crosstalk**

The CID is basically a row column accessed device and it is not surprising that at some level interaction between pixels on the same column or on the same row might appear. The first problem of this nature occurs when a pixel on a row, heavily exposed, affects the output from another pixel on the same row. Figure 7 illustrates what is believed to be a crosstalk mechanism. When the row potential is reduced to zero volts, all the charge in all the row wells is transferred to column wells. A sense amplifier connected to a given column will see not only the charge transferred into that specific well, but also some fraction of the charge transferred at other wells along the row. It is felt that the coupling capacitance between row and column accounts for this. The magnitude of the cross-coupling is about 0.025%/pixel.

The problem can be alleviated by a first order correction in which 0.025% of a summation of pixel amplitudes along a row is subtracted from each pixel. That is: a single constant is determined for a row which is subsequently subtracted from each pixel.

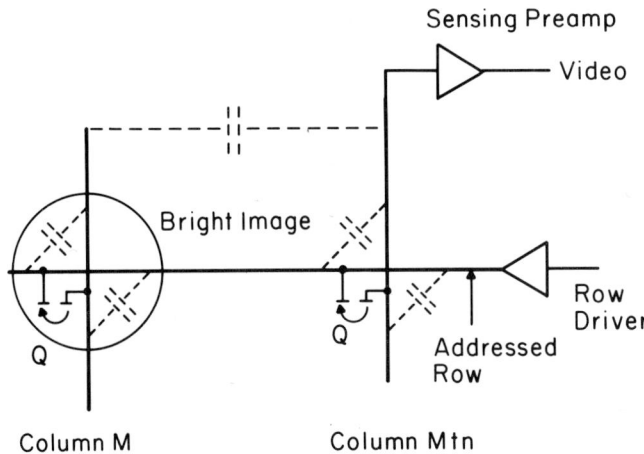

Fig. 7. Possible mechanism which produces horizontal coupling between elements along rows.

III. Applications

The CID was tested on several telescopes on Kitt Peak; Figure 8 is a 15-minute exposure taken with the 4-meter Mayall Telescope with a narrow band filter at the Cassegrain focus. The object is M-57, a classic planetary nebula. The CID was applied to the problem of detecting a faint potassium emission shell around α Orionis. A narrow band (0.2 Å) was used to filter out all but the desired wavelengths and 5-minute integrations were made at the 4-meter coudé focus.

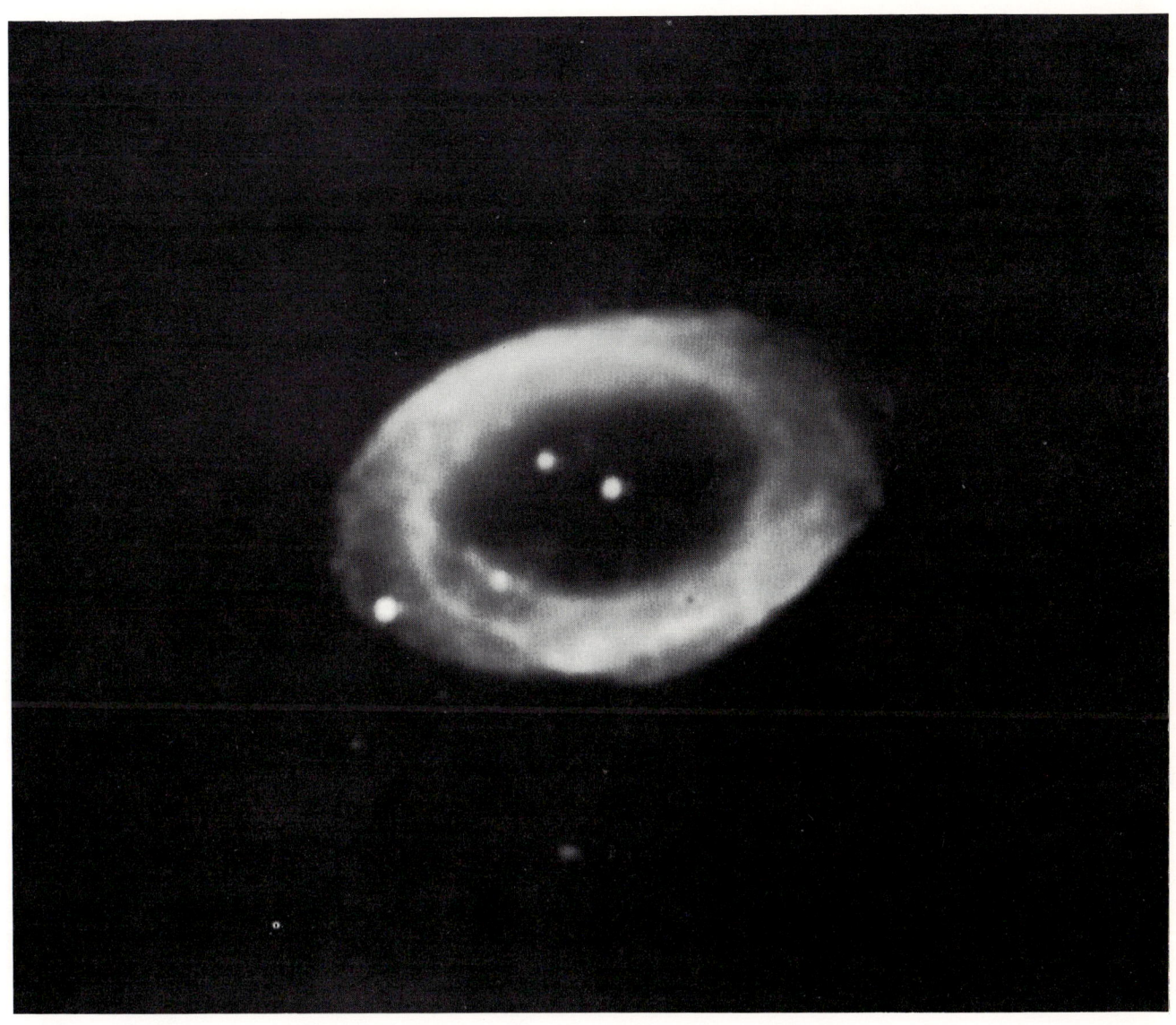

Figure 8. M-57 in Lyra, taken with the CID.
Streaking of central stars is due to the display monitor video amplifier
and does not appear in the original data.

IV. Conclusions

The data given here indicate that the CID is a unique and extremely promising device for use in astronomical applications. The non destructive readout, superb dynamic range, and high spatial quality make this device one of the most useful solid state arrays for astronomical applications today.

During the next nine months three CID cameras will be constructed at Kitt Peak National Observatory. These cameras will exploit the non destructive readout and random access features of the CID and will be used in a variety of solar and stellar applications. Cooling will be achieved through the use of a demountable dewar.

V. References

Michon, G.J. and H.K. Burke, "Charge Injection Imaging," 1973 IEEE International Solid State Circuits Conference, and "Operational Characteristics of CID Imager," 1974 IEEE International Solid State Circuits Conference, published in Proceedings of "Symposium on Charge-Coupled Device Technology for Scientific Imaging Applications," held at the Jet Propulsion Laboratory, March 6-7, 1975.

Richman, Paul, MOS Field-Effect Transistors and Integrated Circuits, John Wiley & Sons, 1973.

Question by Geroge R. Carruthers:

Have CID's been tried in the electron bombarded mode (ICID)?

Answer by R. S. Aikens:

General Electric Company has done some preliminary work in that area. The person to talk to is John Hooker.

Question by Norbert Thonnard:

What is the single frame read-out noise for the CID? Is it the same for destructive or non-destructive readout? What was the amount of charge involved in the pictures that were brought out by repetitive readout?

Answer by R. S. Aikens:

Readout noise is presently around 500 electrons in both modes. It is possible to signal average in non-destructive mode and improve S/N by the square root of the number of frames.

Question by Edgar A. McLean:

What is the linearty of this device?

Answer by R. S. Aikens:

Measurements of signal output vs light input over the operating range give a linearity of $\pm 0.2\%$.

Question by S. H. Koeppen:

Is dry ice cooling sufficient for your purposes?

Answer by R. S. Aikens:

No. Liquid nitrogen is used because at dry ice temperatures, dark current is still a dominant source of noise. Several pixels show unusually high dark current at dry ice temperature.

INTENSIFIED CHARGE COUPLED DEVICES FOR
ULTRA LOW LIGHT LEVEL IMAGING

Stanley Sobieski
NASA, Goddard Space Flight Center
Greenbelt, Maryland 20771

Abstract

The development of high performance solid-state sensors is important for future astronomical space flight missions. Although sensitivity in the ultraviolet is obviously advantageous, the potential for superb imaging afforded by space telescopes recommends a sensor design covering a broad spectral range including the visible and near infrared. Additionally, the sensor must be rugged, reliable, and relatively insensitive to corpuscular radiation. With these considerations in mind, the Laboratory for Optical Astronomy has begun the development of a sensor incorporating a CCD and photocathode for operation in the electron bombarded mode in both a magnetic and an electrostatic focussed, gated configuration. Sufficient gain will be available to provide adequate signal-to-noise for the detection of individual photoelectron events. Initial format size is 100 x 160 pixels with eventual growth to 400 x 400 pixels. The sensor is part of a digital camera system which includes the low level video conditioning electronics, a camera controller, a high speed buffer memory, and digital recording and display electronics. The memory uses CMOS/SOS and has a capacity of 1.6 Mbits with operating rates of 48 Mbits/sec. Individual frames are co-added to provide wide dynamic range and photometric precision better than 1%. A 4-bit video quantization is used to increase the photon counting detection rate before coincidence losses become serious.

Introduction

As one of its objectives, the Laboratory for Optical Astronomy at GSFC is engaged in the development of new sensors to secure astronomical data from above the earth's atmosphere. The purpose of this paper is to describe one program to develop a camera system suitable for either direct imaging or imaging spectroscopy for space missions in the 1980's from the Shuttle or free-flying satellites. A central concern in the selection of the camera sensor, is its capability of being utilized by a variety of instrumentation thereby eliminating the high cost of space-qualifying numbers of special purpose sensors. An example of a candidate mission for the application of the camera is the Large Space Telescope; a 2.4 meter telescope providing diffraction limited imaging performance with sensitivity to wavelengths just shortward of L_α, 1216Å. It is a truism that the performance of such an instrument as for many groundbased telescopes critically depends on the performance of the sensors both in regard to responsivity and to resolution. But in addition to these photometric requirements, the space application imposes additional requirements and conversely constraints regarding size, power, and reliability. Furthermore, since the sensor is subjected to a hostile radiation environment particularly in low earth orbit missions involving passages through the trapped particle radiation belts and their extension, the South Atlantic Anomaly, its design must consider means to reduce the subsequent increased background noise and, in the longer term, the potentially irreversible performance degradation. From an operational viewpoint, the sensor ideally should be such as to minimize the required exposure time to achieve a given signal-to-noise ratio. This is tantamount to maximizing the detective quantum efficiency both by maximizing the quantum efficiency or response and minimizing the internal or system generated noise. Insensitivity to gamma rays and particle radiation by signal discrimination techniques is highly desirable. These requirements appear to be met as far as current work indicates by a solid state camera tube incorporating a Charge Coupled Device.

One of the early suggestions[1] for the concept of a sensor based on the use of CCD operated in an electron bombarded mode was made by E. J. Wampler during a symposium concerned with candidate detectors for the large Space Telescope held in 1972. Signal detection of individual photoelectron events based on electron bombarded Silicon with direct readout had already been successfully demonstrated by the work of Beaver and McIlwain[2] in 1971. During this same time period, the Jet Propulsion Laboratory began the development of a thinned, accumulated CCD at Texas Instruments, Inc. It is configured for back-side illumination by photons; however, it was apparent that electrons could also be used with minimal risk of damage due to X-ray generation within the gate metallization and without incurring excessive energy loss by the electrons in penetrating the surface layers. Thinning to approximately 10μm promised to minimize loss of resolution accompanying excessive diffusion path lengths. In December 1973, the Advanced Sensor Group in the Laboratory for Optical Astronomy began work in cooperation with the Night Vision Laboratory, Ft. Belvoir, capitalizing on the JPL effort to develop a first generation ICCD, intensified charge coupled device, in a proximity focussed configuration utilizing a Bi-alkali photocathode on MgF_2 to achieve wavelength sensitivity in both the visible and ultraviolet. The results of this effort and subsequent second generation sensor activities is documented by Williams[3]. Since that time functional designs for the camera have been undertaken in-house at Goddard with the most recent activity being the commencement of the design and fabrication of the engineering model of a flight camera memory.

Digital Camera Description

The dey component of the digital camera is the ICCD. An ICCD, is a sensor in which photoelectrons ejected from a photocathode are accelerated and focussed within a tube configuration to impinge on the CCD, thereby producing multiple charges according to the total energy imparted by the acceleration voltage applied to the tube. By combining the CCD with a photocathode two important advantages are obtained. First, by selecting the proper combination of photocathode type and substrate material, the wavelength sensitivity range of the camera can be tailored to the application. For example, CsTe on MgF_2 provides sensitivity in the 115 to 300nm wavelength range and thus acts as to discriminate against visible photons which can contaminate faint ultraviolet measurements. Bi-alkali photocathodes have been used in our work because they provide in addition to ultraviolet response, visible response, thereby facilitating laboratory testing. Also there was a reluctance to attempt fabricating tubes with S-20 photocathodes in the early phases to avoid possible contamination problems with the CCD by Sodium, an important constituent of this cathode type. The second advantage is that adjustable gain well in excess of unity can be obtained. Although typical CCD quantum efficiencies can approach 80% in the visible they are unity gain devices and hence, multiple photon events must be recorded to obtain adequate signal-to-noise. Gain then reduces the required exposure time to detect signals above a given threshold.

In the electron bombarded mode of operation, the CCD is used first to provide signal gain, second, to store the resultant image charge pattern and third, to implement the signal readout to serial recording devices. In principle the signal electrons can impinge on the front, metallized side of the CCD but in the process undesirable charging of the SiO_2 layer can occur as well as potential damage by resultant x-rays, and subsequent generation of noisy interface states. Thinning the CCD to approximately 10μm and properly passivating the back surface allows operation with electrons impinging on the back surface and thereby avoiding these aforementioned problems. Since the depletion regions below the electrodes extend approximately 4μ, the undepleted path length for the carriers is approximately 6-8μ which is small compared to the pixel spacing of 23μ which is obtained in the present devices. This results in excellent image resolution following the expected $\sin \chi / \chi$ relation as reported by Collins[4] and Ando[5].

Gain is achieved by the loss of energy by the incident accelerated photoelectron in the creation of electron-hole pairs, one for each 3.6eV loss. Although the backside Silicon surface is treated to minimize the recombination velocity of the minority carries some initial energy loss, on the order of 3 to 5KeV, is not recoverable as gain due to recombinations at the surface.

In order for the ICCD to be able to detect photon events unambiguously sufficient gain must be provided so that the smallest output signal due to a photoelectron event is clearly detectable above all system noise. If in addition, as discussed by Tull[6] the ratio of the dispersion in the gain compared to the gain is small, as it has been shown for the EBS gain process, equal signal weighting for photons results. The various noise components which must be considered are as follows:

1. CCD noise sources such as dark, "fat or slim zero", fast interference states, and bulk, trapping noise - these sources depend on the physics of the Silicon and result either in the injection of fluctuating signal or in the statistical modification of the true signal as it is transferred. Cooling minimizes at least the bulk Si dark noise.

2. Charge transfer efficiency effects - In addition to bulk trapping, different transfer times apply for the transfer of a packet of charge thereby producing a smeared charge distribution in the process. This in turn results in a signal diminution as a function of pixel location within the array. For realizable transfer efficiency of 99.99%, approximately 10% of the signal can be lost by this effect.

3. Electronic noise - The output pre-amplifier produces additional noise with the lower limit being set by the output capacitance, i.e. the kTC noise. CCD's have characteristically $C \sim 2 \times 10^{-13}$ farads thus minimizing this noise while simultaneously allowing for an excellent charge conversion efficiency of 1μV/electron.

By using correlated clamp amplifiers to eliminate the "kTC" noise and cooling to reduce the dark noise one can achieve rms noise values of 50-100 electrons. For signal gains of approximately 4000, it should be possible to achieve signal-to-noise ratios of 10:1 for the detection of 1/4 of a photoelectron, the worst case condition if the charge deposited by a single photoelectron is distributed among four adjacent pixels. This provides better than a 3σ peak level discrimination against detector noise. The overall operating voltage required to achieve this gain is approximately 19kV, a value readily attainable in magnetic and electrostatic focussed tube configurations.

In a single (or multiple) photon counting mode of operation the ICCD must operate with an external memory in which, for the most direct approach, each memory word location corresponds to a distinct pixel. The camera memory currently in design is a static one which utilizes CMOS/SOS devices with 1024 bits per chip. The design depends only upon components which can be space qualified. The memory is organized into 8 parallel sections, each having its own 16 bit Adder, although a design involving a single Adder is also being considered. This multiplexing arrangement permits the ICCD to operate at 6Mpixel/sec data output rates while allowing greater than 1μsec R-M-W cycle times for the memory devices. Twofold redundancy and hence, increased reliability is achieved by having the CCD operate in half sections into separate, but latched, camera memories. The store size per camera memory is approximately 8×10^5 bits. The functional block diagram of a system designed for flight use is shown in Figure 1.

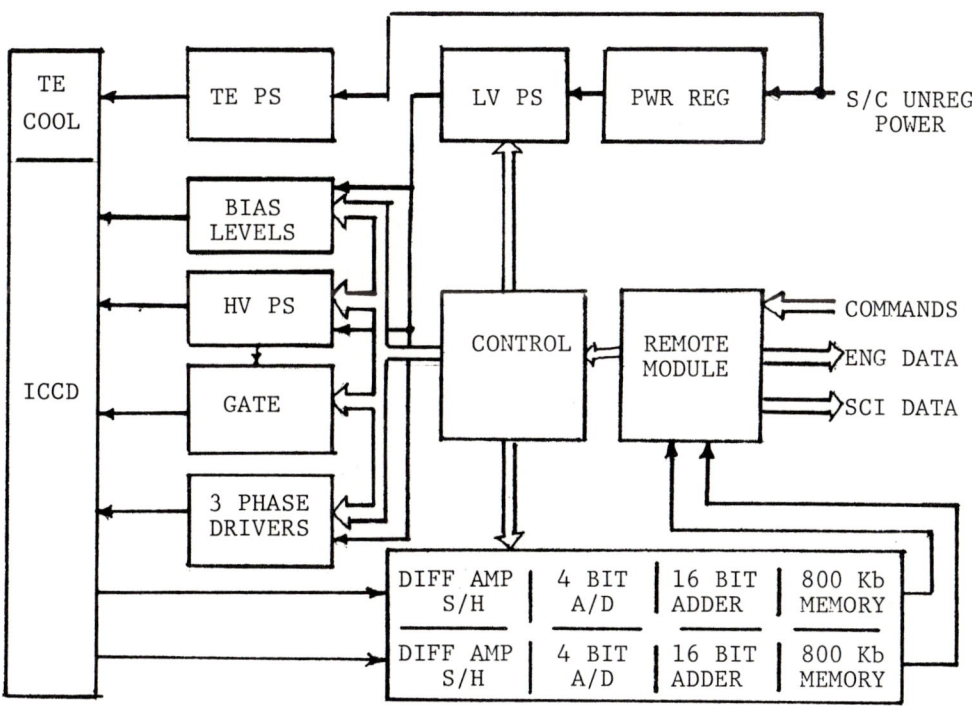

Figure 1. Digital Camera System

During an integration period, the photoelectrons corresponding to the input photon image produce an image signal within the CCD. During readout, while the detector is gated off in order to eliminate image smear, the signal is converted into a serial data stream by the inherent organization of the CCD, digitized in a 4 bit A/D converter and stored in a high speed buffer memory. Subsequent frames of data are co-added to any stored digital data. This sequence is repeated until sufficient signal statistics are achieved. The integration time should be sufficiently short to avoid pulse coincidence errors but the total frame time must be sufficiently long to allow a complete read-modify-write operation of the data into memory.

If the two ICCD sections are readout simultaneously at a 6Mpix/sec rate, a frame is read out in 10msec. For a total frame time of 25msec, i.e. a 60% duty cycle for signal integration, a continuous photoelectron rate of 5×10^5/frame/sec can be handled with negligible coincidence losses. Signal quantization at a 4 bit level extends this by additional factors of 4, again remembering that the least significant bit corresponds to the charge associated with a 1/4 photoelectron. Of course, if an unfavorable duty cycle is allowed, say for bright sources, the tube can be gated on for very short integration periods, 10μsec being a practical lower limit, thereby extending the photometric operating range without incurring pulse coincidences losses. As demonstrated by a number of investigators working with Silicon vidicons and SIT camera tubes, several orders of magnitude reduction in Silicon dark current can be achieved with moderate cooling. Thus, the digital camera being described when cooled to -40°C or -60°C by thermoelectric devices will be capable of photon noise limited performance operating in an analog mode, i.e. utilizing the CCD to store charge for up to 300 seconds with less than 10% dark signal accumulated. In this case the gain can be substantially reduced to recover single frame dynamic range and the output signal would be quantized with a greater resolution than the 4 bits used in the photon counting mode. It is this mode of operation which will be most compatible with the considerably larger format CCD's, e.g. the 400 x 400 pixel device currently under development at JPL. As a long range plan, the ICCD, in these and larger formats, will be improved replacements for the SIT camera tube, affording better signal-to-noise and, in a proximity focussed configuration, ultra-stable, distortionless image quantification.

ICCD Tube Configuration

At the outset of this program, the primary concerns regarding (a) CCD tube header design, (b) compatibility of the CCD and typical tube fabrication schedules and (c) easily controlled investigation of the EBS gain process appeared to be best addressed utilizing a proximity focus tube configuration and the 100 x 160 pixel CCD which was well into development by JPL. Although this phase of the program described is still underway by Williams[7], the second phase of the effort has begun. It consists of the development, again at Texas Instruments, of a 400 x 250 pixel thinned CCD organized into two independent but latched sections of 400 x 125 pixels and the development at ITT Corporation of a magnetically focussed tube capable of operation to 25Kv and utilizing a mesh gating electrode in proximity to the photocathode. The primary objectives for this second phase are to investigate the problems of operating in a photon counting mode and to utilize the breadboard camera system to observe at groundbased telescopes. However, for future space flight application, an electrostatic focussed tube configuration, an EFICCD is being seriously considered and some preliminary design work on representative camera housings have been completed. The EFICCD is an attractive configuration for several reasons. As shown by Cromwell and Dyvig[8] the configuration because of the intrinsic interior shielding is less susceptible to ion noise. As important, the configuration provides the required signal gating without the use of a scanning coil system or lossy, mesh electrode. Because of its compact design compared for example to that required for a magnetic focussed tube with its large focus coil, the sensor can be operated at smaller off-axis angles in a folded optical system, thereby simplifying the optics design. Until recently, the main disadvantage of this tube configuration was its curved electronic focal plane, a problem easily remedied for visible sensitive tubes by the use of figured fibre optic faceplates. The pentode design developed by NVL, however, provides for planar input and output and hence the MgF_2 window needed to obtain UV sensitivity, can be accommodated. The schematic design of EFICCD is shown in Figure 2.

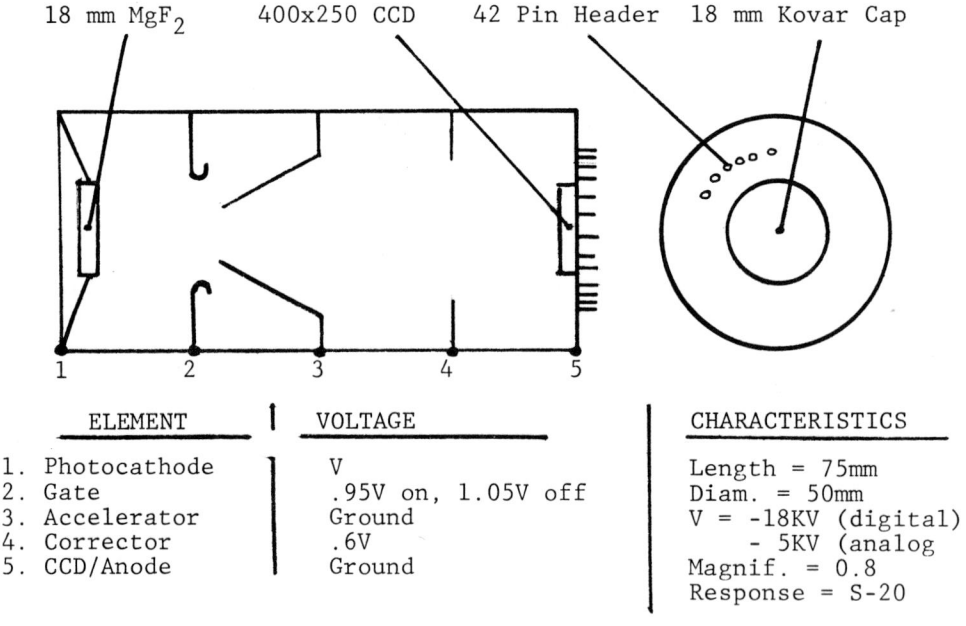

ELEMENT	VOLTAGE	CHARACTERISTICS
1. Photocathode	V	Length = 75mm
2. Gate	.95V on, 1.05V off	Diam. = 50mm
3. Accelerator	Ground	V = -18KV (digital)
4. Corrector	.6V	- 5KV (analog
5. CCD/Anode	Ground	Magnif. = 0.8
		Response = S-20

Figure 2. Electrostatic Focus Pentode Tube

Summary

Study and preliminary design efforts as well as laboratory tests at the component level, indicate the feasibility of developing a solid state camera system for ultra-low light level applications for groundbased observing and for use in space flight missions. Such a system using CCD's once developed, should be reliable and considerably more stable than systems incorporating electron read gun sensors. Cost effectiveness follows from the generalized format of the basic sensor and its ability to operate in either a digital or analog mode with only changes in the external control logic.

Acknowledgement

I wish to thank my colleagues in the Advance Sensor Group, in the Laboratory for Optical Astronomy, J. Shannon, C. Sturgell, and J. Williams, and W. Miller and J. Yagelowich of the Spacecraft Data Management Branch, for their efforts and contributions to this work. The support and encouragement of forward looking managers at NASA Headquarters and the endorsement of an unnamed peer review group is gratefully acknowledged.

References

1. Advanced Electro-optical Imaging Techniques. NASA SP-338, edited by S. Sobieski and E. J. Wampler, p. 62, 1973.

2. Beaver, E. A. and McIlwain, C. E., Rev. Sci. Instr. 42, 1321, 1971.

3. Williams, J. T. "Testing Results on Intensified Charge Coupled Devices". This symposium, 1976.

4. Collins, D. R., Roberts, C. G., Chan, W. W., Rhimes, W. C., Barton, J. B., and Sobieski, S., "Development of a CCD for Ultraviolet Imaging Using a CCD Photocathode Combination", Proceedings of the Symposium on CCD Technology for Scientific Imaging Applications. JPL SP 43-21, p. 163, 1975.

5. Ando, K. J. "MTF and Point Spread Function for a Large Area CCD Imager", Ibid p. 192.

6. Tull, R. G., Choisser, J. P., and Snow, E. H. "Self Scanned Digicon", Applied Optics 14, p. 1182, 1975.

7. Williams, J. T., "The ICCD as a Photon Counting Imager", Proceedings of the AAS Symposium - Space Shuttle Missions of the 80's, paper No. 75-208, 1975.

8. Cromwell, R. H. and Dyrig, R. R., "Laboratory Evaluation of Eleven Image Intensifiers" Tech. Report #81, Optical Sciences Center, U. of Arizona, Aug. 1973

Question by K. Kubierschky:

Is the ICCD which you are developing to be considered a NASA standard?

Answer by S. Sobieski:

It is not. This work is based on the particular interests of the Laboratory for Optical Astronomy. However, its design is such that it will be space qualifiable and also be versatile enough to operate in a wide range of imaging applications at low light levels.

Question by K. Kubierschky:

What are the details of the sensor format?

Answer by S. Sobieski:

The sensor format is organized into two sections of 400 pixels/line by 125 lines each, for a total of 10^5 discrete pixels. Each pixel is approximately 23μm square. The two sections each have their separate output pre-charge amplifiers and separate clock lines and can be read out simultaneously. However, the two sections can be operated in tandem by externally controlling the clock signals thereby allowing operation in the event of an amplifier failure for either section.

TEST RESULTS ON INTENSIFIED
CHARGE COUPLED DEVICES

Jack T. Williams
NASA, Goddard Space Flight Center
Greenbelt, Maryland 20771

Abstract

CCD imaging arrays have been evaluated for application in a photon counting astronomical sensor. The results obtained show that CCD devices presently being manufactured exhibit noise corresponding to less than 100 rms electrons and gains of approximately 10^3 at an accelerating voltage of 10kV and greater than 5×10^3 at 25kV. These results demonstrate that the CCD devices, when operated in an intensified mode, can be used for photon counting. In addition, the measured charge transfer efficiencies are in excess of 0.9999, signal saturation levels are as high as 9.9×10^5 electrons per pixel and peak quantum efficiencies are greater than 75%. With these characteristics, the arrays will make excellent analog detectors surpassing commercial vidicon performance in small format applications. Successful proximity focussed intensified charge coupled devices (ICCD) have been fabricated, although problems still exist with contamination and possible structural failures in the CCD's during tube manufacture.

Introduction

Since the introduction of the CCD concept in 1969[1], progress in design and fabrication techniques has proceeded at an extremely rapid pace. As a result, CCD imagers are today finding many applications and one of the more important is the recording of low light level astronomical sources. In this area, solid state detectors are very advantageous and can fulfill a longstanding requirement for a two dimensional imager that is both phtoometrically and geometrically accurate while consuming reasonable amounts of space and power. Therefore, a program of test and evaluation of CCD imagers and tube processing techniques has been undertaken with the goal of developing an intensified charge coupled device for photon counting astronomical applications[2,3].

Intensified Charge Coupled Device

Operations

The ICCD consists of a CCD[4] operating in the electron bombardment mode and being used as the output of an image intensifier/converter. Thus, the ICCD combines the advantages of solid state TV readout with the gain and wavelength selectivity possible in an image intensifier/converter and will have performance superior to a CCD alone[5].

In the electron bombarded mode of operation, the CCD is used first to provide signal gain, second to store the resultant image charge pattern, and third to implement the signal readout to serial recording devices.

Gain is achieved by the loss of energy of the incident accelerated photoelectron in the creation of electron-hole pairs, one for each 3.6eV loss. Although the silicon surface is treated to minimize the recombination velocity of the minority carriers, some initial energy loss (of the order of 3 to 5KeV) is not recoverable as gain due to recombinations at the surface. An ICCD is a sensor in which photoelectrons ejected from a photocathode are accelerated and focussed within a tube configuration and impinge on the CCD chip, producing multiple charges according to the total energy imparted to the photoelectrons by the acceleration voltage applied to the tube.

CCD Selection

In order for the ICCD to be able to detect photon events unambiguously, sufficient gain must be provided so that the smallest output signal due to a photoelectron event is clearly detectable above all system noise. Thus, low noise is the key factor to be considered in selection of CCD's for ICCD applications. Normally, electrons can impinge on the front, metalized side of the CCD, but in the process undesirable charging of the SiO_2 layer can occur as well as potential damage by resultant x-rays, and subsequent generation of noisy interface states. These problems may be avoided by thinning the CCD to approximately $10\mu m$ and properly passivating the back surface allow operation with electrons impinging on the back surface. CCD noise sources such as "fat zero" and surface trapping noise at high frequency can be avoided or minimized by utilizing buried channel CCD's which in addition, tend to operate better at low signal levels. Bulk dark noise in the silicon can be kept to a minimal by cooling. Use of carefully designated correlated clamping[7] or distributed floating gate[8] amplifiers can further reduce electronic noise. As a result, tests show that a thinned, backside illuminated, buried channel CCD that is cooled and uses a correlated clamping amplifier can realize rms noise values as low as 50-100 electrons[9]. With noise values this low and with the gain expected from an ICCD, acceptable signal-to-noise levels for photon counting are achievable.

TEST RESULTS

Lab Equipment and Test Programs

As shown in Table 1 the laboratory is equipped to do most testing that will be required on CCD's and ICCD's.

Table 1. Laboratory Equipment

9 Track Incremental Digital Recorder
ASR 33 Computer Control Teletype
System Power Supplies
8000 Channel PHA & MCA
Solid State Memories
PEP Image Storage Device
ICCD Program Controller
Image Display Devices
Nova II Mini-Computer
Mico-Projector
Low Level Light Source

The purpose of this equipment is to provide a real time diagnostic and test capability for large format photon counting imager systems. In addition to the equipment in Table 1, various bandpass, color and neutral density optical filters are available for the imaging tests, as are calibrated diodes and photomultipliers for absolute sensitivity measurements.

The test program being established will completely characterize the device under test with respect to the parameters listed in Table 2.

Table 2. Test Parameters

CCD Quantum Efficiency
Photocathode Efficiency
Charge Transfer Efficiency
Modulation Transfer Function
Dark Current Noise
Pre-Amp Noise
Spatial Variation in Dark Noise
Spatial Variation of Gain Transfer Characteristics
Saturation Level
Uniformity of Response
Geometrical Distortions
Temperature Characteristics

CCD Status

Goddard Space Flight Center has had tested 15 CCD devices of the type described here which were fabricated under contracts with Texas Instruments Corporation and the Army Night Vision Laboratory. These devices were mounted on ceramic tube headers[10] specially designed for incorporation into tubes. The tests which were conducted emphasized the CCD characteristics in an electron bombarded mode of operation, and Table 3 is a summary

Table 3. Measured CCD Characteristics

Parameter		Typical Value	Best Value
Quantum	0.45 m	30%	75%
Efficiency	0.80 m	58%	71%
(15 Devices)	1.1 m	5%	8%
Resolution for 50% SWAR*	unfiltered	19.1 Lp/mm	19.8 Lp/mm
(7 Devices)	0.4 m	14.5 Lp/mm	15.0 Lp/mm
Noise	na/cm^2 (15 Devices)	26	4.2
	RMS electron (7 Devices)	237	78
Saturation level	(7 Devices)	7.2×10^5 elect.	9.9×10^5 elect.
EBS Gain at 10kV	(7 Devices)	824	1050
Charge Transfer Efficiency	(15 Devices)	0.9976	0.9999

*Square Wave Amplitude Response

of the results of those evaluations. The EBS gain was determined with an electron scanning microscope. During the testing period the technique for measuring the exremely low noise values was gradually improved and the values quoted are subject to large experimental errors. More recent refined measurements show noise values characteristically less than 100 rms electrons.

The dynamic range of these buried channel devices is relatively large, with high quality devices having values of $6-8 \times 10^3$ at unity gain. It can be seen that they will be excellent analog detectors with quantum efficiency and dynamic range superior to conventional vidicons. It should be noted that the important EBS gain characteristic is conveniently correlated with the blue photo response of the CCD, thereby facilitating simple testing for devices selection. This correlation reflects the similarity of high energy photons and electrons as they interact with the passivated surface of the device and it clearly indicates the importance of this technology area if optimum results are to be achieved.

The two outstanding problems with the Texas Instrument CCD's are blooming under high input signal conditions and non-planarity of the thinned membrane, particularly for the larger format CCD's. Night Vision Laboratory and Jet Propulsion Laboratory are studying these problems respectively, and Goddard Space Flight Center expects to benefit from their results.

Additional testing of the CCD's has been performed at Texas Instrument, Corp. regarding the response to a point source of electrons. An electron beam was scanned across several pixels to determine the point spread function, that is, to assess the magnitude of lateral diffusion within the bulk silicon. The peak response was between 93 and 97%, across and parallel to the channel stops. A 50% response was measured with the beam positioned precisely at the junction of two pixels. These results[11] indicate negligible diffusion of signal from one pixel to the next and, if the CCD is operated in a digital mode where small analog signal shifts can be discounted, that an MTF approaching that given by the pixel pitch can be expected.

Although Table 3 shows the electron gain of the CCD for 10kV, all chips were tested to determine their EBS gain at voltages ranging to 25kV. It was found that gains in excess of 5000 were observed which is adequate for the intended application in a photon counting ICCD. The measurement of the gain dispersion characteristics, nearly as critical for the projected mode of operation as the gain itself, remains to be done.

However, measurements simulating an EBS operational model[12] showed narrowly peaked gain characteristics and for this reason a favorable result is anticipated. Of more concern is the non-uniformity of the thinned CCD surface passivation which could result in macroscopic gain variations deleterious to signal quantization at a fractional photoelectron level. These tests have shown that CCD's of this type and exhibiting the typical performance parameters in Table 3 are suitable for use in a photon counting ICCD for astronomical application.

ICCD Status

Although it was expected that a diode tube ICCD would probably not have adequate gain to allow photon counting directly, such devices were built first in order to obtain quickly a flight qualified device, having high reliability and well defined characterictics. In addition, the diode tube provides the essential preliminary information required for development of a photon counting tube, that is information on the compatibility of the photocathode and the thinned CCD within the vacuum envelope and on those tube processing techniques which may adversely affect the CCD. ITT Fort Wayne was selected to manufacture a proximity focus design ICCD, and five Texas Instrument CCD's were delivered to ITT for incorporation into tubes. ICCD No. 1 was delivered to Goddard Space Flight Center in July 1975. Tests on this device revealed that the only measured CCD parameter to be affected by the tube processing was a slight increase in dark current. The ICCD was operated in the EBS mode at voltages up to 12.5kV and limiting resolution, defined by the proximity tube characteristic of 7 lp/mm, was measured. Before further tests could be completed an accidental over-voltage was applied to the CCD destroying the on-chip amplifier and making the device unuseable.

The next two tube starts resulted in very large increases in the CCD dark currents as an apparent result of tube processing. Analyses of these devices and an additional two test devices indicated that photocathode materials had contaminated the CCD's. There also appears to be a problem in the method of bonding leads to the CCD. During tube processing, where large temperature cycles are encountered, the bonds are apparently weakened causing poor contact. In one case the bond became completely loose. The temperature cycle problem is further exemplified in tests at Goddard Space Flight Center where four CCD's were operated in a cooled mode and exhibited erratic behavior, with the CCD putting out no signal at temperatures below 0°C and then coming back on as the temperature rises. This phenomenon has apparently not been observed before and no clear understanding of the mechanisms involved exists although the most likely source is the CCD-header interface which includes the lead bonding problem mentioned above. Detailed analysis of these problems is currently being conducted at Goddard Space Flight Center and Texas Instrument.

ITT has begun processing on the remaining diode tubes after modification of their processing station to allow better protection of the CCD, more accurate temperature monitoring, and in situ evaluation of the CCD during processing.

In addition, ITT was fabricated a magnetically focussed tube having a phosphor output to verify the design of a photon counting ICCD. Further work on this ICCD will proceed immediately upon completion of the diode work. These magnetically focussed tubes will incorporate a fine mesh electrode immediately behind the photocathode to provide electronic gating. They will operate at 20kV in order to have sufficient gain for pulse counting and will contain MgF_2 windows and Bi-alkali photocathodes.

Conclusion

It has been demonstrated by evaluation of actual devices that CCD imagers can be fabricated which possess the necessary performance characteristics for use in an ICCD. This performance has been verified by the manufacture of an operating proximity focussed ICCD. Manufacturing problems do exist, however, and detailed diagnostic procedures are needed during fabrication in order to identify the causes and cures of contamination and structural failures in the CCD's. A true photon counting ICCD has now been designed and prototype production has begun.

References

1. Boyle, W. S. and Smith, G. E. (1970) "Charge Coupled Semiconductor Devices", Bell Systems Technical Journal, 49, 587-593.

2. Currie, D. G. "On a Photon-counting Array Using the Fairchild CCD-201", Proceedings, Symposium on CCD Techniques for Scientific Imaging Applications, JPL, 15 June 1975, pp. 80-90.

3. Williams, J. T. "The Intensified Charge Coupled Device as a Photon Counting Imager", AAS Conference on Space Shuttle Missions of the '80's, Aug. 1975, paper No. AAS 75-208.

4. Sequin, C. H. and Tompsett, M. F. "Charge Transfer Devices", Academic Press, 1975

5. Barton, J. B., Curry, J. J., Collins, D. R., "Performance Analysis of EBS-CCD Imaging Tubes/Status of ICCD Development", CCD Application Conference Proceedings, 1975 at Naval Electronics Laboratory Center.

6. Kim, C-K and Dysk, R. H. (1973),"Low Light Level Imaging with Buried Channel Charge Coupled Devices", Proceedings, IEEE, 61, 1146-1147.

7. Barton, J. B., "Development of Electron Bambraded Silicon (EBS) Charge Coupled Device (CCD) Imager", Final Technical Report by Texas Instruments, Incorporated on Night Vision Laboratory Contract No. DAAK02-74-C-0359, (May 1975), pp. 11-13.

8. Wen, D. D., et al., 1975,"A Distributed Floating - Gate Amplifier in Charge Coupled Devices,"ISSCC Philadelphia Digest of Technical Papers, pp. 24-25.

9. Barton, J. B. (see Reference 6 above) pp. 119-165.

10. Barton, J. B. (see Reference 6 above) pp. 17-22.

11. Collins, et al., 1975, "Development of a CCD for Ultraviolet Imaging Using a CCD Photocathode Combination", Proceedings, Symposium on CCD Technology, JPL, SP 43-21.

12. Beaver, E. 1974, "LST Final Instrument Definition Faint Object Spectrograph", Appendix D, p. 96.

A PHOTON COUNTING ARRAY PHOTOMETER

Douglas G. Currie
Department of Physics & Astronomy
University of Maryland
College Park, Maryland

and

John P. Choisser
Electronic Vision Company
Science Applications, Inc.
San Diego, California

Abstract

The operation and performance of an array photometer with the ability to discriminate on single photoelectrons will be discussed. The University of Maryland Array Photometer is a system which consists of a 100 by 100 array of channels which count single photoelectrons. The photosensor is an Intensified Charge Coupled Device using a Fairchild CCD201 which was fabricated by the Electronic Vision Company. This system has demonstrated a noise level which permits single photoelectron discrimination, high scan rate (400 frames/sec), and low lag. The system and its performance will be discussed. The UMAP system has been successfully operated in the photon counting mode on a 36-inch telescope and on a 48-inch telescope. Data from these observations will be discussed, to illustrate the system sensitivity and the spacial resolution which is better defined by the size of an individual pixel element on the CCD.

I. Introduction

The University of Maryland Array Photometer is a photon-counting, rapid-scanning array of channels, each of which is capable of single photoelectron discrimination. The photodetector of this system is an internally Intensified Charge Coupled Device fabricated by the Electronic Vision Company and built around a Fairchild CCD201/202. Several of these Intensified Charge Coupled Devices (ICCD) have been fabricated by the Electronic Vision Company (a division of Science Applications, Inc.) and then tested at the University of Maryland. These tests have been conducted primarily with the Single Scan Data Recording System with some preliminary tests conducted with the full University of Maryland Array Photometer.

General Properties of the Array Photometer

The University of Maryland Array Photometer (UMAP) system, using an ICCD, is an ultra-sensitive TV-type camera system which distinguishes individual photoelectrons in each of 10,000 channels. This is in the form of an array of one hundred by one hundred channels. The time resolution in each channel is normally shorter than three milliseconds, which is the time required to scan one frame. The frame time and thus the time resolution may be much smaller if a smaller subarray is scanned. The array is sensitive to the incident light for essentially all of the observational period (i.e., greater than 98% of the time) and the dynamic range for the basic system is very large, approaching 10^8.

Internally Intensified Charge Coupled Device

The basic photodetector for the UMAP system is the ICCD. In this device, an incident photon is converted to a photoelectron at the photocathode. In the current tubes, the photocathode has response similar to an S-20. The photoelectron is accelerated, focused, and enters the region of active silicon in the Charge Coupled Device, where ionization produces a charge packet. The scanning circuitry of the CCD provides the parallel-to-serial conversion of the data, and on-chip preamplification of the video signal.

History and Current Status

The development of the UMAP electronics and the SSDRS was started late in 1973, and a preliminary set of electron bombardment tests of a CCD were performed at the Electronic Vision Company in August of 1974 using the preliminary form of the SSDRS. The initial contract for tube fabrication from the University of Maryland to the Electronic Vision Company was let in October of 1974. ICCD-1 was received for testing at the University of Maryland in June of 1975. This tube proved to be gassy after an initial series of tests were completed. ICCD-3 was received at the University of Maryland in September of 1975, and has shown no problems with gas. ICCD-4 was just recently received in February of 1976. The results of earlier tests have already been reported[1], and this report will primarily focus on the results obtained with ICCD-3.

The overall system is currently operating at a noise level which permits single photoelectron discrimination and permits the formation of images with discriminated single photoelectrons. The system has been operated in this fashion on a variety of astronomical telescopes. Although the UMAP system is not yet in full operation, and the results are quite preliminary, the various aspects of the system which have been carefully tested are in accord with the early theoretical predictions.[2]

II. Requirements for the Array Photometer

The development effort described in this report has been motivated by two different applications at the University of Maryland. These applications, although they are both astrophysical in nature, generate a significantly different set of requirements on the array photometer.

Amplitude Interferometer Requirements

The primary application of the UMAP system is to provide the basic photosensor for a new instrument, the Multi-Aperture Amplitude Interferometer (MAAI). This unique new system which is a multi-channel version of a similar Amplitude Interferometer has been used in an astronomical observation program over the last four years[3,4,5,6]. When used on a large telescope, it will yield better than diffraction-limited image information through the earth's atmosphere. It will have an angular resolution somewhat better than the currently operating instrument of i.e., 0.010 arc-seconds when used on the 200-inch telescope on Palomar Mountain. The precision of the measurements with the current instrument are about 0.002 arc-seconds. This will be greatly improved with the new MAAI, which has a data rate which is larger by a factor of about 7,000. This instrument utilizes all of the incoming light on the telescope primary. Basically, this application requires an array of individual photosensors, each of which has:

1. The ability to discriminate reliably among zero, one, two, and more than two photoelectrons per pixel per scan,

2. Very low lag, i.e., very little memory from one frame to the next frame,

3. Minimal cross talk between spacial channels (or pixels) and high geometric stability.

The requirements on the overall array are:

1. At least 100 by 100 elements and the ability to sub-scan a smaller array,

2. The ability to complete the scan of a full frame in a few milliseconds, and subframes at a rate of much faster than the basic 400 frames a second.

Direct Imaging Camera Requirements

The other application in which the Array Photometer will be used consists of a Direct Imaging Camera in the focal plane of a large telescope. This will both operate in conjunction with the MAAI, as well as in a stand-alone mode. For this application, most of the above requirements are important, but the imaging application has the additional requirements of:

1. Very large dynamic range,

2. Very low blooming.

Some of the electronic operating parameters of the system will be significantly different for the two applications.

III. Method of ICCD Operation

In this section, we briefly describe the theory of operation and the detailed design of the ICCD. Both of these areas have been previously described in more detail in the literature[2,7].

Theory of Operation of the ICCD

The incident photons are received on a photocathode, where they are converted into photoelectrons. As indicated in Figure 1, these photoelectrons are accelerated to an energy of about 15 Kev and electrostatically focused onto the front surface of a Fairchild Charge Coupled Device (CCD201, or for the current and future ICCD's, the CCD202).

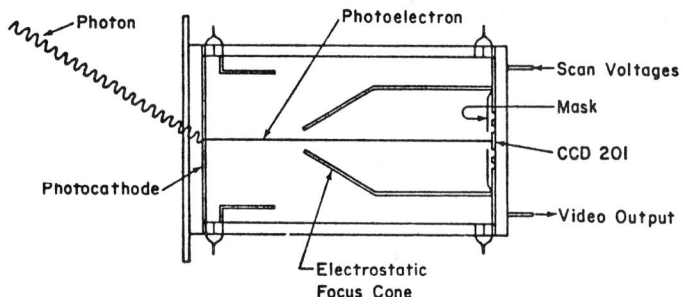

Schematic Diagram of Intensified Charge Coupled Device
Figure 1

The photoelectron penetrates the overlying layers of circuitry (where some energy is lost) and creates a packet of charge at one of the photosites by ionization of the silicon. This accumulation of individual packets of charge continues until the end of the integration cycle, when the charge packets are transferred from the photosites to the transfer registers. The transfer registers carry the charge from the two-dimension array of photosensitive sites through a parallel-to-serial conversion and to the on-chip preamplifier. These transfer registers operate independently of the photosites and at the same time that the photosites are integrating the electrons produced by ionization for the next frame. Thus the transfers or "scanning" take place during the integration period, and the array is sensitive more than 98% of the time.

The transfer registers are protected by a layer of aluminum on the front surface of the CCD and the accelerating voltage is chosen so that those bombarding electrons which fall on the transfer sites do not reach the active silicon. Thus one loses a factor of two in effective quantum efficiency, but one does not have any ionization or "interline noise" in the transfer registers.

Depending upon what region of the photosite the electron enters, a packet of about 1,770 or 2,040 ionization electrons is produced for each photoelectron. Thus, with a noise of 300 electrons in the on-chip preamplifier, one may easily discriminate on single photoelectrons.

ICCD Design

The design of the ICCD incorporates a number of features which enhance the ability to discriminate at the single photoelectron level. A special CCD201, on which most of the top layer of SiO_2 (the scratch protection) was not added, was mounted by the Fairchild Corporation on a ceramic header. This header, designed by the Electronic Vision Company, provides the required electrical contacts for the CCD in a circular pattern of pins. Thus it may be mounted with a socket into the camera head developed for the ICCD. The header also provides the support for the electrostatic focus cone and the mask which exposes the active array and protects the preamplifier and other circuitry from damage by accelerated photoelectrons. The chip is bonded directly to the ceramic header so the CCD may be readily cooled using a probe which contacts the header directly below the CCD in an area which is clear of pins. This cooling may be required to reduce the thermally generated dark current on the chip.

IV. University of Maryland Array Photometer System

Since the UMAP system described in this section has been developed primarily as the photosensor for the Multi-Aperture Amplitude Interferometer, several aspects of the system are especially oriented toward this goal. In particular, there are several requirements on the system beyond those general requirements discussed in Section II which are related to operation at the Cassegrain or prime focus of a large telescope. These are:

1. Operation over a wide temperature range,

2. Photometer unit which is light weight, mechanically and optically rugged, and operable at all orientations,

3. High immunity to external RFI,

4. Proper operation with main electronics subsystems located 100 meters from photometer unit, i.e.,

a. a minimum of cross talk between the cable transmitted signals,

b. proper clock sequencing to handle propogation delays of several cycles,

5. Equipment packaging to permit the use of air freight in an "assembled" configuration.

The following discussion will center on the use of the ICCD as the photosensor for the UMAP system. However, the UMAP system is also designed to use bare CCD's as infrared sensors with a few minor additions. The use of the UMAP system within the MAAI involves a rather complex overall system. Thus to simplify the following discussion, we shall describe the use of the UMAP system with the Direct Imaging Camera. This will use part of the MAAI system, but the additional capabilities and complications will be either ignored or mentioned only in passing in this discussion.

Block Diagram of System

The block diagram of the UMAP system, as it will be used in the "Direct Imaging Camera" mode, is shown in Figure 2.

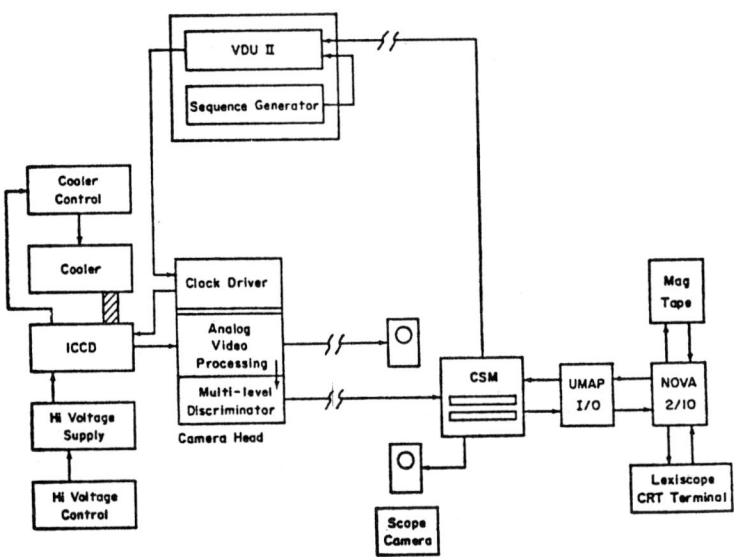

Block Diagram of UMAP System as Used in the
Direct Imaging Camera Mode
Figure 2

We shall now proceed to describe the overall operation of the UMAP system. Depending on the particular requirements of the observations, the scan rate, the format, and the size of the scan of the array is chosen by the operator. The parameters which describe this subarray are typed into the Lexiscope CRT interface. The NOVA mini-computer then generates the detailed instructions which are required for scanning the CCD in this particular fashion. The NOVA cannot operate at the speed required to directly control the CCD, so these commands are transferred and stored as special control words in the Circulating Semiconductor Memory. The Circulating Semiconductor Memory (CSM) is a memory device designed and fabricated for this application. It has five tracks (only one of which is used in the Direct Imaging Camera mode) containing 12,288 words, each of which is 16 bits. The CSM, which operates at a data rate up to five megahertz, transmits these individual commands to the Video Drive Unit (VDU) which is located on the telescope near the photometer unit. The Video Drive Unit generates coded pulse signals which, at the camera head, are converted to high current pulses to drive the CCD. The CCD is cooled by a thermoelectric device to reduce the variations of the dark current across the chip to a value which is small compared to the random capactive input noise on the on-chip preamplifier. This temperature is typically between 0°C and -20°C. The video output from the CCD, which contains the data, is brought to an electrically isolated portion of the camera head, separate from the chamber which contains the drive circuitry. This configuration reduces the pick-up of coherent noise by the sensitive video electronics. The video signal is amplified and the signal divided for two types of processing. On the one hand, the charge from each photosite is discriminated at various preset voltage levels for use in a photon-counting mode. On the other hand, an A/D conversion is performed on the other portion of the signal. This permits the operation of the CCD in a linear; rather than photon counting mode. The information, in digital form, is transmitted from the telescope down to the Video Processing Unit. The Video Processing Unit has a very simple function for the Direct Imaging Camera. However, in the MAAI, it performs the on-line computations which are required to transform the data into a structure in which one may add successive frames of data. From the Video Processing Unit, the data stream proceeds to the Circulating Semiconductor Memory, where this data is added, pixel by pixel, to the data of the previous scans. For permanent storage of the data, the NOVA minicomputer slows down the CSM in a predetermined sequence, transfers a block of data from the CSM to the NOVA core storage, sets the registers in this block in the CSM to zero, and accelerates the CSM to the proper rate and reinitializes the CCD. The data is then transfered from the NOVA core storage to digital magnetic tape.

System Performance

Let us now consider the predicted performance of the overall system. The intrinsic dynamic range for the UMAP system is extremely large (i.e., greater than 10^8). Since the overall system is not yet operating at this level, these predictions will be based on data from several sources, some of which will be discussed in later sections. However, in most applications, practical aspects of the application will prohibit the use of such a large dynamic range. For example, when the system is used for direct images of faint objects with a wide spectral range on a large telescope, the effective dynamic range is defined by various parameters of the overall UMAP/telescope system and of the atmosphere. Thus to make a realistic projection for a given observational situation, we will consider the overall configuration which was to be used for quasar observations on the 200-inch telescope at Palomar Mountain in January of 1976. The photometer unit was installed at the prime focus with Ross corrector lens system and a set of spectral filters. However, to achieve the maximum performance at high dynamic range, none of these optics would be used. For the photocathode response, we use the measured quantum efficiency of the "S-20" surface of ICCD-1. For this illustrative calculation, we shall express the results in terms of the image brightness expressed in stellar magnitudes per square arc-second. Thus we primarily address the study of an extended object, like a distant galaxy. We note, however, that if the seeing disc is slightly larger than one arc-second, the ordinate and absissa of Figure 3 may also be correctly interpreted as the visual magnitude of unresolved source. Thus we have an analysis which permits a condensed presentation to both the observation of a quasi-stellar or unresolved object (expressed in stellar magnitudes) and to an extended object (i.e., a faint galaxy) where the surface brightness is expressed in stellar magnitudes per arc-second. A more complete discussion, including the dependence upon seeing, will be presented elsewhere. Thus in Figure 3, we have

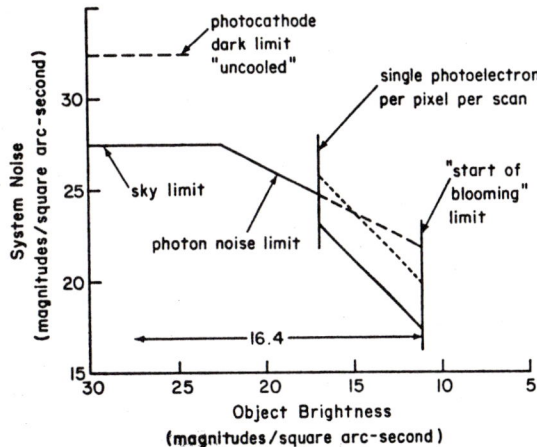

Limiting system noise of UMAP system, where the absissa describes the surface brightness expressed in magnitudes per square arc-second and the ordinate describes the equivalent brightness of the "noise" or uncertainty in an observation of a pixel expressed in stellar magnitudes.

Figure 3

For an object with a brightness between 16.9 and 22.5, this uncertainty is due to the Poisson counting statistics in the signal from the object with an assumed observation interval of ten minutes. This also assumes a brightness of the moon-less night sky at Palomar of 22.5 magnitudes. For objects fainter than 22.5 magnitudes, the dominant noise source is the Poisson statistics in the night sky. In the range from 16.9 to 11.2, the noise may be either Poisson or systematic as discussed later. We have presumed an "uncooled" photocathode. The plate scale is about 11"/mm, giving a field of 33" by 44". The UMAP system will discriminate among zero, one, two, and more photoelectrons[2]. Thus one may have a mean count rather close to the frame rate of 400 frames a second, and still have a reasonably small correction for four and more counts. We shall nominally assume a mean rate of 400 photoelectrons per second per pixel for the maximum rate which is still in the single photon counting rate. For the faint limit of the dynamic range, we obtain brightness of 27.5 magnitudes which is equal to the statistical Poisson noise in the night sky, i.e., we have here a signal-to-noise ratio of one. Remaining in the photon counting mode, we then have a dynamic range of 10.6 magnitudes or 17,400.

However, there is no blooming until at least several hundred photoelectrons per pixel per scan are accumulated. Thus if we were to record data on the SSDRS at the same time as the photon counting data is recorded in the CSM, we could obtain an additional factor of, conservatively 200 in the dynamic range. The current camera heads permit this but it has not been implemented on the rest of the equipment. Thus we see from the figure that the dynamic range for this configuration is 16.4 stellar magnitudes or about 3.6×10^6.

Thus this mode of operation includes the use of the data from regions bright enough to require operation in the analog region of the CCD. For this analog measurement, we have presumed that the limit may not be the photon counting statistics, but the accuracy to which the systematic problems may be corrected. For the solid line in Figure 3, we have presumed that the systematic errors can be corrected to a level at which we have already found that we can perform this correction during some preliminary tests at the University of Maryland, (i.e., 0.3%). The dotted line indicates a correction at a level of 0.05%, and we note that Fairchild has published data claiming this may be done by 0.025%.

This projected performance makes the UMAP system particularly interesting for observing objects which have a very large dynamic range over a relatively small field. Since the response of each individual pixel is independent at this signal level, one can expect operation here with very little blooming. For the configuration of the telescope and the UMAP system described in this section, the effect of the insensitive aluminum transfer registers on the photometric accuracy for observing unresolved objects is less than 0.05%.

In the operation mentioned, the limiting magnitude is defined by the statistical noise in the night sky background. However, in the absence of the night sky, there are two effects which will limit the ultimate dynamic range. The first of these is the emission of thermal electrons from the photocathode in the absence of light (photocathode dark current) and the other is caused by large random noise excursions tripping the discriminator (electronic dark current).

The evaluation of the level of electronic dark rate makes several assumptions.

1. The "random noise"[2] of the CCD has a standard deviation of 300 electrons,

2. The distribution of this random noise is indeed Guassian to the required level. Preliminary data on this point will be discussed in a later section of this paper,

3. The chip architecture, and the resultant pulse height distribution have the form which has been described earlier[2],

4. The operating point is set at 1320 electrons to yield a collection efficiency of 93%.

This results in a limiting magnitude of $m_v = 32.4$.

"The "uncooled" dark current limit presumes that the system is operating at an about "observatory" temperature of 50°F which we have found to be representative of the dome temperature. This calculation is based upon dark current rates which we have observed in the field and at the University of Maryland on similar devices with S-20 surfaces. This is in accord with published S-20 data.[8] One then obtains a limiting magnitude of 32.2, or about one photoelectron in ten minutes which is indicated in Figure 3. The addition of a procedure for cooling the photocathode would reduce this noise source, but then the electronic dark noise would immediately dominate.

For applications in which one is not limited by sky noise (i.e., imaging with narrow filters or spectroscopic work), we would have a photon counting dynamic range of 1.3×10^6 and a total dynamic range of 2.6×10^8. However, these results are extremely sensitive to the assumption made on the electronic dark current.

VI. Laboratory Performance

In this section we consider various tests and evaluations which have been conducted in the laboratory.

Single Scan Data Recording System

In order to evaluate the performance of the various portions of the system prior to the full operation of the electronics of the UMAP system, a test system, the Single Scan Data Recording System[9] (SSDRS) was designed, fabricated, and put in operation in May of 1974. A block diagram of the SSDRS is shown in Figure 4.

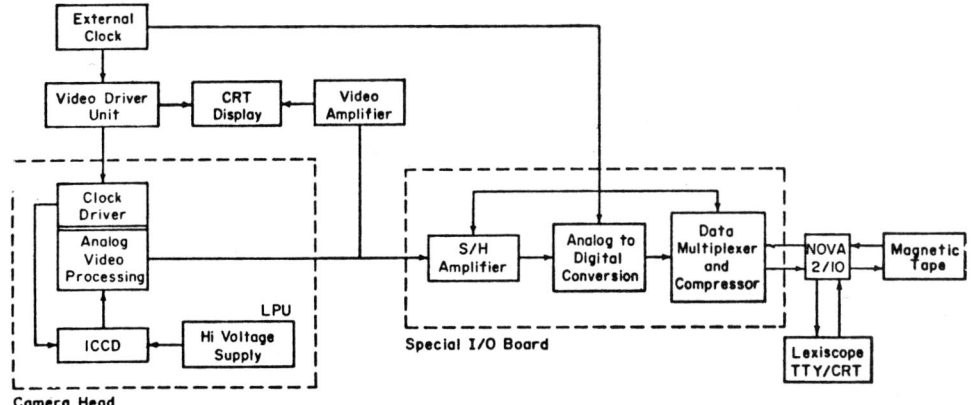

Block Diagram of the Single Scan Data Recording System
Figure 4

The basic mode of operation is to perform an A/D conversion for each pixel at a data rate which may be as high as four megahertz. A single frame of data words is then stored in the NOVA core. This stored frame of data is written onto digital magnetic tape. The rate at which frames may be recorded is limited by the magnetic tape writing speed to about three frames per second. Most of the laboratory and telescope data has been collected with this system. In several cases there appears to be some noise pickup which may dominate the effects of the CCD and camera head. In addition, we have recently discovered a one pixel shift in the recording of the data which has led to an apparent increase in the noise. Recent measurements, after fixing the jitter, on ICCD-4 which has a CCD202, indicate a random noise of about 300 electrons.

CCD System Noise Level

In this section we consider effects or "noise" which would degrade the system performance. The first of these effects to be considered here is the "random noise", followed by a consideration of the "fixed pattern noise".

We now consider the apparent random noise which one observes at the CCD. This has been measured in various laboratories to have a typical value of about 300 electrons.

In order to evaluate our data for this purpose, we shall use a pulse height distribution over many elements of the array. The pulse height distributions are formed by first recording ten frames of dark, which are then averaged. One then subtracts from a single illuminated frame, pixel by pixel, the above averaged dark frame and performs a pulse height analysis of the remaining array. Usually this is done over a central subarray, to reduce computer costs and to eliminate edge effects. This is repeated for ten frames and the ten pulse height distributions which are produced are summed to produce the indicated pulse height distribution. This may be displayed directly (Figure 7) or we may normalize the peak of the pulse height distribution to unity and plot the upper side of this curve as seen in Figure 5.

The random noise of the CCD is indicated in Figure 5 and is seen to be about 1.1 A/D Converter Units (ADU). The effect of turning on the high voltage is indicated by the pulse height distribution denoted with o. The difference is not significant, since differences of this magnitude occur for runs separated by an interval of time. We expect the standard deviation of this noise to decrease when the discriminators mounted in the camera head are used and indeed during a series of tests using the full UMAP system, in December, this occured although these results were not recorded in the computer.

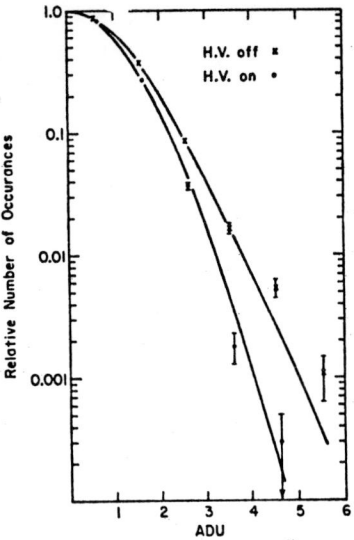
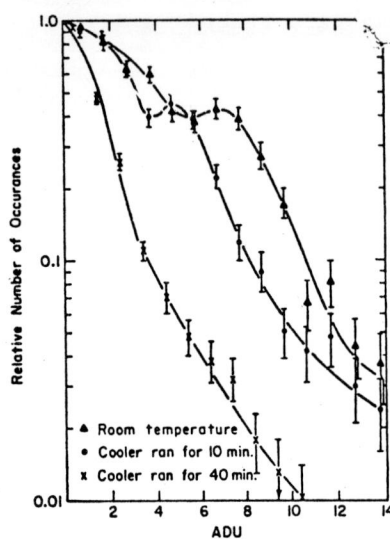

Pulse Height Distribution of Random Noise for ICCD-3. The ADU are each 0.124 millivolts at the CCD.
Figure 5

Pulse Height Distributions of Thermal Leakage at Three Different Temperatures
Figure 6

The other source of noise is the "fixed pattern noise" or the spacial variations of the thermal leakage current in the silicon. The two types of noise are independent and combined in a simple manner, so they may be studied separately. The fixed pattern noise is evaluated by averaging ten frames (to reduce the effect of random noise) and then performing a pulse height distribution on the averaged frame. At room temperature and at a scan rate of 0.5 MHz, these variations in dark current are expressed in Figure 6 and are unacceptably large. The variation becomes much smaller and quite acceptable as the CCD is cooled. When the CCD is operated at 4 MHz, the width of the peak should be reduced by another factor of eight.

Single Photoelectron Pulse Height Distribution

We now wish to evaluate the magnitude of the signal from a single photoelectron. In order to do this, the photocathode is illuminated with a uniform intensity and the recorded data is used to obtain the single photoelectron pulse height distribution. This has been done with various voltages and various light intensities. Three of these pulse height distributions are indicated in Figure 7.

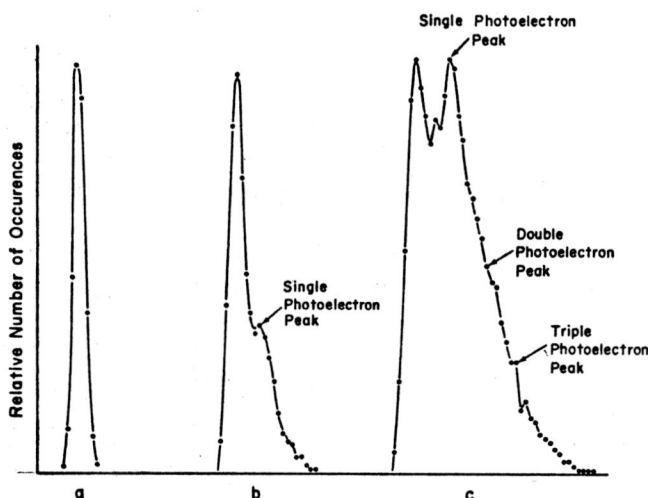

Pulse height distribution with 15 KV, 15 KV, and 18 KV respectively and with no illumination, illumination, and illumination respectively.
Figure 7

Figure 7a illustrates the random noise of the CCD with no light input. This shows a relatively narrow peak with a standard description of 1.2 ADU. In Figure 7b, the photocathode is uniformly illuminated with an accelerating voltage of 15 KV. The peak due to the single photoelectrons can be seen. The peak due to zero photoelectrons is somewhat enlarged to about 1.5 ADU, and the mean magnitude of the single photoelectron signal is 5.3 ADU. In Figure 7c, where the accelerating voltage has been increased to 18 Kv, the photoelectron peak is further broadened to about 2.0 ADU. This may be due to some penetration occuring at this (above nominal) accelerating voltage. The position of the single photoelectron peak (with respect to the ICCD photoelectron peak) is 7.5 ADU. Finally, the width of the single photoelectron peak is 3.0 ADU. In order to compare these values with the predictions, we must determine the gain of the on-chip preamplifier. The nominal value[10] is 1.5 Volts/picocoulombs, however, this may have wide variations[10]. Thus to determine the value for this present chip, we use the change in accelerating voltage and the change in peak position. This implies $380e^-$/ADU or the gain of 2.0 Volts/picocoulomb, and implies the measured signal at 14.4 KV (which was calculated in an earlier work[2]) would be $1780e^-$ compared with the theoretical value of $1910e^-$, well within the accuracy of this determination and knowledge of the various layers on the chip surface. The additional width of the singles peak is also in agreement with predictions[2]. That is, in Figure 7c, the singles peak is 1.0 ADU wider than the zero photoelectron peak and the prediction is that the additional width is 0.7 ADU.

The behavior of an individual pixel in the ICCD is comparable to the performance of a conventional photomultiplier. Using the data expressed in Figure 7c, and setting the discriminator at a level to obtain a collection efficiency of 84%, one has a dark current rate of one count per second, which is a rather respectable performance, even at this preliminary stage.

Spacial Resolution

The spacial resolution of the electrostatically focused ICCD-3 was evaluated. using a microprojector to place a small spot of light (~7μm) on the photocathode. In the central region (out to a radius of about 1.3 mm in the 3mm by 4mm format), essentially all of the light fell within a single picture element. In this region, the diameter of the electron image of the incident light was less than 20 micrometers. This was determined by moving the incident light spot until the electron image fell on the aluminum transfer register between two photosites which has a width of 20 micrometers. No response was seen in either of the adjoining photosites where the sensitivity of this determination was at least 10% of the peak response. Thus in this region, the MTF of the UMAP system is basically the MTF of an array of discrete pixels. However, further out from the center the electron image spread to cover more than a single pixel.

VII. Field Operation

Operational tests using the Single Scan Data Recording System with the bare CCD's and the ICCD's have now been conducted on a variety of telescopes. The purpose of these tests have been to:

1. evaluate field problems (low temperature, transportation, "fool-proof operation")

2. determine the sensitivity of the system to external interference (TV and FM)

3. evaluate the system sensitivity

4. evaluate single photoelectron detection capability.

Some of these tests were conducted on the 36-inch telescope which is operated by the Laboratory for Optical Astronomy and the 48-inch telescope operated by the Laser Technology Branch. Both of these groups are within the Goddard Space Flight Center and are located at the Goddard Optical Research Facility (GORF). In addition, the photometer system was operated at the 60-inch and 100-inch telescopes at Mt. Wilson, and preliminary tests were also conducted on the 200-inch telescope at Palomar Mountain, which are operated by the Hale Observatories. A detailed list of these observations appears in Table I.

Field Configuration of System

The system configuration which has been used in these tests consists of the Single Scan Data Recording System. Although the CCD has not been cooled, the ambient temperature has ranged, on various observational runs, between 0°C and 20°C. A number of different objects have been observed in order to calibrate the sensitivity and to evaluate the general performance.

TELESCOPE	FOCUS	DATE	OBJECT	PHOTO-SENSOR	PLATE SCALE ARC-SEC/PIXEL	COMMENTS
100-inch Mt. Wilson	Cassegrain	8 July 1975	g Her, etc.	CCD	0".178	System operation Test of TVI sensitivity
Mt. Wilson	50 mm lens**	7 July 1975	Antares	CCD	146"	Sensitivity test & camera calibration
60-inch Mt. Wilson	50 mm lens	19 July 1975	Jupiter	CCD	146"	System operation Test of TVI sensitivity
36-inch GORF*	135 mm lens	24 Oct. 1975	Jupiter	ICCD-3	54"	Determination of the system QE
Sorrento Valley ††	135 mm lens	28 Oct. 1975	Pleiades	ICCD-3	54"	Blooming and resolution measured
48-inch GORF	Coude	November 1975	o UMa	ICCD-3	0".197	Determination of the system QE
36-inch GORF	Cassegrain	16 Nov. 1975	Jupiter, γ Ari, 30 Tau, Landolt 095-0206	ICCD-3	".56	Determination of the system QE
36-inch GORF	Cassegrain		Landolt 095-0206	ICCD-3	".56	System Sensitivity
36-inch GORF	Cassegrain	19 Nov. 1975	30 Ari, Saturn Trapezium	ICCD-3	".56	Evaluation of imaging
" "	"	"	Saturn		1".00	Evaluation of imaging
100-inch Mt. Wilson	Cassegrain	28-31 Dec. 1975	ξ UMa, α Per, 14 Aur, 39 Leo	ICCD-3	0".178	Star calibration of sensitivity and effects on TVI determined.
200-inch Palomar Mtn.	Prime Focus	8-10 Jan. 1976	Star	(ICCD-3)	0".39	†

* Goddard Optical Research Facility, Beltsville, Maryland
** 50 mm lens and 135 mm lens are camera lenses
†† Electronic Vision Co., Sorrento Valley, San Diego, California

† Due to power supply accident no observations performed with ICCD, however, some observations were made with a simulated ICCD. Alignment and operational procedures developed.

Table I
Operation of UMAP System on Astronomical Telescopes

Telescope Mounting of Photometer Unit

The photometer unit (which is about 4" by 5" by 6") was mounted in a different fashion at each of the telescopes. On the 36-inch telescope at GORF, this was mounted at the Cassegrain focus by means of an existing adapter, a 4" by 5" plate holder. The operation using the 48-inch telescope at GORF was at the Coude focus and the photometer unit was tripod mounted within the Coude room. The mounting on the 100-inch telescope was at an auxiliary focus at the Naismith Cassegrain position. This was formed with a bending mirror attached to the Amplitude Interferometry Mounting Adapter. The mounting on the 200-inch was at the prime focus using the SIT camera bracket while the Amplitude Interferometer was mounted at the Cassegrain focus. This permitted rapid change over between the two instruments (approximately 15 minutes).

Results of Telescope Operation

We first consider Figure 8, which shows the random noise when the system is operated on the telescope at the Goddard Space Flight Center.

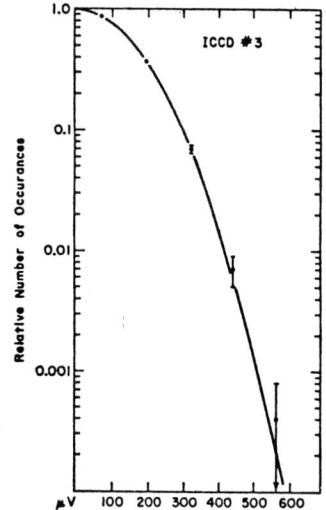

Pulse Height Distribution of Random Noise Measured on 36-inch Telescope at GORF

Figure 8

For the calculation of the dynamic range, it was assumed that the statistical behavior of the random noise was Gaussian over 4.5 orders of magnitude. In Figure 8, the Gaussian curve, fitted to the three hightst points, is indicated by the solid line. Although this data does not cover the full range of five decades, this data is Gaussian, within the statistical error bars of the measurement. Even at Mt. Wilson, where 14 Megawatts of TV and FM power are emitted within one mile of the dome, and where the analog signal sent along 150 feet of cable, the apparent random noise was not excessive and the monitor display was quite acceptable. This apparent random noise due to external pickup should not be a problem when the discrimination is done within the camera head.

System Performance

We now discuss the results which were achieved with the SSDRS operating on the various telescopes. The general purpose of the stellar observations has been to calibrate the sensitivity of the system. As indicated in the table, the stellar observations have been conducted on the 36-inch, the 48-inch, and the 100-inch telescopes. Generally, these tests have indicated a system response within a factor of two or three of the predicted system response. A more precise comparison is difficult, since the quantum efficiency of the photocathode dropped by a factor of ten from the values measured during tube fabrication (about 20% at peak) to the values measured at the end of the tests. The reason for this decrease is being investigated. However, the system seems to behave as predicted using the value of the quantum efficiency measured after the test. This assumes the loss was due to an event early in the tube lifetime. In addition to the stellar images, direct images of several objects were recorded. In particular, Saturn was observed with a neutral density filter to reduce the intensity by 100. This resulted in a multi-photoelectron signal per pixel per scan. The set of four stars which compose the Trapezium was observed with neutral density filters which reduced the light level by factors of 10 and 100. The first set of data had about ten photoelectrons per pixel per scan, while the second set of data on the Trapezium had about one photoelectron per pixel per scan, at the brightest region.

The Trapezium data was then discriminated in the computer. That is, the dark frames were averaged and subtracted from each frame with the image on it. Then each time a pixel had a signal as large as one photoelectron, a total frame was incremented by one unit. This "software discrimination" was applied to a set of 100 frames of data, and yielded a reasonable image of the configuration. During exposure to normal (low) light intensity, there has been no obvious cumulative damage to the CCD by the accelerated electrons, either due to a lack of damage inflicted or due to self-annealing between exposures. When the tube (ICCD-3) was exposed to high levels of light, there has been obvious increases in dark current which have apparently not self-annealed. Similar affects due to high level illumination have already been reported[1]. However, these regions of increased dark current become negligible in magnitude when the CCD is cooled.

Conculsions

An array photometer using an ICCD as the photodetector has been designed, developed, and fabricated. It is currently being tested in the laboratory and on the telescope. The performance in the tests up to this time has been generally in accord with predictions.[2]. The system has performed imaging on the telescope with discriminated single photoelectrons. The ICCD has operated reliably in the laboratory and in the field for five months. A test of the full system is expected shortly.

Acknowledgements

Many people and groups have aided this effort. We very much appreciate the generous amounts of telescope time which have been made available by the Hale Observatories and and by Goddard Space Flight Center, both for the ICCD work, and for the Amplitude Interferometry. At the University of Maryland we would especially like to thank Robert Braunstein, who has been responsible for the major effort is obtaining the operating data on the UMAP system and analyzing this data. The electronics of both the UMAP system and the SSDRS has been brought into reality by John Giganti and the Electronics Shop at the University of Maryland. Most of the development and programming of the NOVA and Eclipse systems has been done by Al Buennagel. The programming and rather extensive computations which have been required for the data analysis have been performed by S. Kaisler, A. Blumenthal, and L. Bleau of our group. This was done on the UNIVAC 1108 of the University of Maryland Computer Science Center which is supported in part by NASA Grant NGR 398. We would like to thank Jim McPherson and Peter Bertling for their work on the ICCD at Electronic Vision Company.

Various different aspects of the work on tube fabrication and system development have been supported by grants and contracts to the University of Maryland from the National Science Foundation (MPS-74-05890-001), National Aeronautics and Space Administration (NGR-21-002-444), Office of Naval Research (N00014-75-C-0343), and to Science Applications, Inc. from USAF Space and Missile Systems Organization (F04701-75-C-0068).

Reference

1. D. G. Currie, AN INTENSIFIED CHARGE COUPLED DEVICE FOR EXTREMELY LOW LIGHT LEVEL OPERATION, presented at the 1975 International Conference on the Application of Charge Coupled Devices, sponsored by the Naval Electronics Laboratory Center, 29, October 1975.

2. D. G. Currie, ON A PHOTON COUNTING ARRAY USING THE FAIRCHILD CCD201, presented at the symposium on Charge Coupled Device Technology for Scientific Imaging Applications, 6 March 1975, p. 80, JPL SP 43-21. This appears in somewhat extended form in University of Maryland Technical Report #75-082.

3. D. G. Currie, ON A DETECTION SCHEME FOR AN AMPLITUDE INTERFEROMETER NAS-NRC, Woods Hole Summer Study on Synthetic Aperture Optics, Vol. II, p. 35, 1968.

4. D. G. Currie, ON THE ATMOSPHERIC PROPERTIES AFFECTING AN AMPLITUDE INTERFEROMETER, NAS-NRC woods Hole Summer Study on Synthetic Aperture Optics, Vol. II, p. 79, 1968.

5. D. G. Currie, S. L. Knapp, and K. M. Liewer, FOUR STELLAR-DIAMETER MEASUREMENTS BY A NEW TECHNIQUE: AMPLITUDE INTERFEROMETRY, The Astrophysical Journal, Vol. 187, p. 1, January 1974.

6. S. L. Knapp, D. G. Currie, and K. M. Liewer, ON THE EFFECITVE TEMPERATURE OF α HERCULIS A, Astrophysical Journal V 198, 561, 1975, University of Maryland Technical Report #75-006, July 1974.

7. J. P. Choisser, EXPERIMENTS ON THE USE OF CCD'S TO DETECT PHOTOELECTRON IMAGES presented at Symposium on Charge Coupled Device Technology for Scientific Imaging Applications, 6 March 1975, p. 150, JPL SP 43-21.

8. LARGE SPACE TELESCOPE, PHASE A, FINAL REPORT.

9. L. A. Buennagel, D. G. Currie, R. Braunstein, A PRELIMINARY DESCRIPTION OF THE SINGLE SCAN DATA RECORDING SYSTEM, University of Maryland Technical Report #76-038.

10 R. H. Dyck, private communication.

LOW LIGHT LEVEL IMAGING DEVICES FOR THE MIDDLE ULTRAVIOLET

G. Carruthers, J. Kervitsky, G. Hicks, C. Opal
E. O. Hulburt Center for Space Research
Naval Research Laboratory
Washington, D. C. 20375

Abstract

Electronic imaging devices based on the magnetically-focused, internal-optic Schmidt image converter concept, previously used at NRL for the vacuum ultraviolet, have been developed with cesium telluride photocathodes for use in the middle ultraviolet (2000-3000 A). These devices are intended primarily for flame and mid-UV source observations, but also have applications to astronomy and to planetary atmosphere studies. Three versions of these devices have been constructed and tested: (1) a single-stage image converter with a phosphor/fiber optic output, for film recording or coupling to a low-light-level television camera tube, (2) an image converter incorporating a microchannel intensifier stage with proximity focusing of the output onto a phosphor screen, and (3) same as (2) except with magnetic focusing from the microchannel plate output onto the phosphor screen. The latter two devices can be used with direct viewing of the phosphor output, or with lens coupling to a film camera or television camera tube. They can also be fiber-optically coupled to a CCD array. In each case, single-photoelectron events are detectable. Cesium telluride photocathodes appear to be significantly more stable in the presence of small leaks and outgassing than are cesium antimonide photocathodes. Therefore, electrographic recording should be easier with the Cs_2Te photocathodes, and preliminary investigations are being initiated.

Introduction

The middle ultraviolet wavelength range is generally defined as the interval 2000-3000 A. For comparison, the far ultraviolet (or vacuum ultraviolet) is the range below 2000 A, and the near ultraviolet is the 3000-4000 A range. In photoelectric detectors and photoelectronic imaging devices, these ranges are distinguished by three basically different photocathode materials which are most useful therein: for the far ultraviolet, alkali halide photocathodes (such as CsI and KBr) are used, for the middle ultraviolet, tellurium-based photocathodes (such as Cs_2Te) are applicable, and for the near ultraviolet and visible, antimony-based photocathodes (such as Cs_3Sb and Na_2KSb) are generally used. In other respects, imaging devices for the three wavelength ranges are quite similar; the only major difference in the complete imaging system is that refractive lenses are more difficult to use in the middle UV and generally cannot be used in the far UV. Therefore, all-reflective and Schmidt optical systems are the most applicable in these wavelength ranges.

Schmidt Image Converters

Electrographic cameras based on the use of a front-surface (opaque) alkali-halide photocathode at the focus of a Schmidt optical system have been in use for far-ultraviolet imagery at the Naval Research Laboratory for the last ten years[1-3] (see Fig.1). These cameras have been used primarily for far-ultraviolet astronomy from sounding rockets[4-6], but were also used on the Apollo 16[7] and Skylab 4[8] missions. Devices of this type have also been built with phosphor/fiber optic output, and with microchannel plate amplifiers followed by either electrographic film[9] or phosphor screen[10].

A distinguishing feature of devices operating in the far ultraviolet is that the alkali halide photocathodes used are not harmed by exposure to dry air, hence unbaked, demountable image tubes of vented construction can be used. However, with photocathodes sensitive to wavelengths longward of 2000 A, the imaging device must be evacuated and baked before photocathode processing; thereafter ultrahigh vacuum conditions must be maintained or else the photocathode will be irreversibly degraded.

We have recently begun to develop imaging devices for wavelengths longward of 2000 A based on the opaque photocathode-Schmidt optical system concept which we have been using in the far ultraviolet. Photocathodes of Cs_3Sb and Cs_2Te have been used, and three versions of these devices have been constructed and tested: (1) a single-stage image converter with a phosphor/fiber optic output, for film recording or coupling to a low-light-level television camera tube (Fig. 2), (2) an image converter incorporating a microchannel intensifier stage with proximity focusing of the output onto a phosphor screen, and (3) same as (2) but with magnetic focusing from the microchannel plate output onto the phosphor screen (Fig. 3).

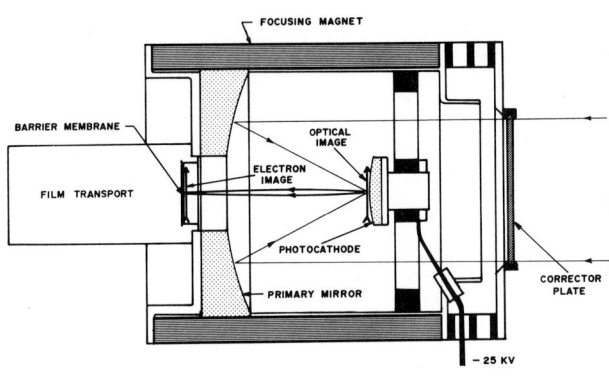

Fig. 1. Electrographic Schmidt camera used with opaque alkali halide photocathodes for vacuum ultraviolet imagery and spectrography.

The devices constructed for use with middle-ultraviolet and visible sensitive photocathodes utilize Schmidt optics with 75 mm corrector plate aperture, a 125 mm diameter spherical primary mirror, and a focal length of 112 mm (f/1.5). Image diameter is 30 mm (15° field of view). The corrector plate is ultraviolet-grade fused silica, separate from the input window of the tube (Corning 9741 UV-transmitting glass, sealed to Koval which is in turn welded into the tube structure). The tube body is constructed of stainless steel with an 8-inch Varian flange assembly at the front end (demountable for tube reprocessing). The tubes are bakeable to 350°C. Photocathodes are presently prepared by first evaporating the tellurium or antimony onto the nichrome-coated fused silica substrate, using an evaporator which is retracted through a 3/8" OHFC copper tubulation (which is then pinched off), followed by evaporation of the cesium using channels permanently emplaced in the tube. The vacuum system line (also 3/8" copper) is also pinched off after completion of processing, but a 2 lit/sec appendage pump permanently mounted on the tube is in continuous operation to maintain the tube vacuum after pinchoff.

The overall quantum efficiencies of the image converters at 2537 A are measured by comparison with a calibrated standard Rb_2Te photodiode obtained from EMR Photoelectric, Inc. At present, the quantum efficiencies obtained using the present system of processing are considerably lower (1 to 5% at 2537 A) than should be ultimately obtainable (~25%).[11] However, we are preparing a

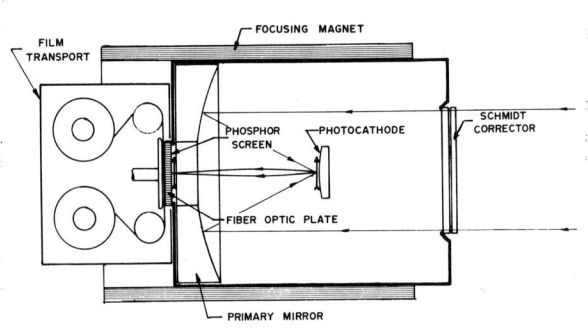

Fig. 2. Schmidt image converter with phosphor/fiber optic output for use with opaque photocathodes sensitive longward of 2000 A.

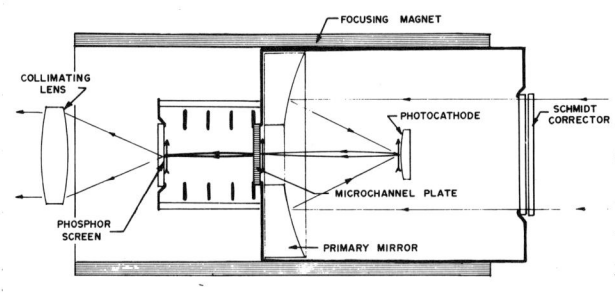

Fig. 3. Schmidt image converter as in Fig. 2 but with microchannel intensifier stage and lens-coupled phosphor output.

separate photocathode processing and testing device which we will use to improve our processing techniques. Fig. 4 compares typical quantum yield curves of various opaque photocathodes and of an S-20 photocathode in the UV-visible wavelength range.

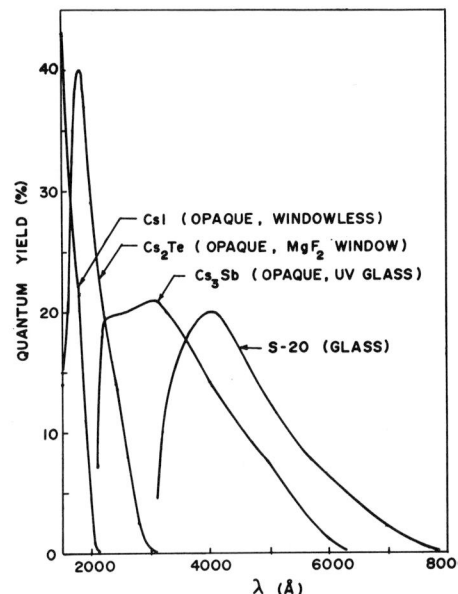

Fig. 4. Quantum efficiency vs. wavelength of opaque photocathodes useful in the UV-visible wavelength range, compared with a typical S-20 semitransparent photocathodes.

The microchannel plates were obtained from Galileo Electro-Optics Corp., either as an integral channel plate plus P-20 phosphor/fiber-optic unit (for the proximity focused device) or channel plate only (magnetically focused device). In the latter case, a P-4 phosphor on glass sealed into a Kovar ring was used.

The microchannel-intensified units can be used with direct viewing of the phosphor output, or with lens coupling to a film camera or low-light-level television camera tube. When operated with microchannel plate gains of a few hundred, single-photoelectron scintillations are readily visible on the output phosphor. Hence, with film recording or LLLTV, the recorded signal is photoelectron-noise limited.

Essentially equivalent performance with the microchannel-intensified devices can be obtained by the use with a fiber-optic phosphor output, coupled to a vidicon or (more interesting) a CCD or CID array. Use of the diode arrays would result in a much more compact system than one using a conventional television camera tube, and without the difficult problem of matching the magnetic fields required by the image converter and by (or not wanted by!) the television camera tube. The only drawback is that the diode arrays are not yet available with as large a picture format (or number of picture elements) as are the conventional camera tubes.

Laboratory Tests

Our present program at NRL is directed to the development of "solar blind" imaging devices for the 2000-3000 A wavelength range, the principal application being the detection of flames and mid-UV sources under the low-background conditions that apply at the earth's surface (due to absorption of solar radiation below 3000 A in the ozone layer). To better define the signal-to-background ratio that would be achievable in practice, we have tested Schmidt image converters with Cs_2Te photocathodes by using them to obtain images of flames in room light (Fig. 5) and outdoors in sunlight. For these tests, an auxiliary interference filter (half-transmission bandwidth 250 A, centered at 2750 A) obtained from Baird-Atomic, Inc. was used to improve the rejection of wavelengths longward of 3000 A. Also, spectra of hydrogen/oxygen, propane/oxygen, and acetylene/oxygen flames have been taken (without filter) using a collimator and plane grating (Fig. 6). Bands of the OH radical, and a continuum (attributed to the reaction of $H + OH \rightarrow H_2O + h\nu$) are prominent in the middle ultraviolet wavelength range. Fig. 7 shows a device, in field-test configuration, of the unintensified phosphor/fiber optic output type shown in Fig. 2. This camera was used with contact recording on Kodak Tri-X 70 mm film to produce the images shown in Figs. 5 and 6.

Tests are presently underway using the microchannel-intensified tubes. These are being used with lens coupling to a 35 mm Nikon camera, a 16 mm movie camera, and an 18 mm SIT television camera. The latter tube is electrostatically focused, and is adversely affected by the magnetic field of the image converter focusing magnet, even with the phosphor-to-photocathode spacing of 13 inches. Therefore, a solenoid coil was fabricated and placed around the front end of the SIT tube, and operated so as to produce a reverse field which nearly cancels the image converter field in the vicinity of the SIT tube photocathode.

The SIT-and movie-camera versions of this system are intended primarily for applications involving rapid time variations. The still-camera version is most useful for static scenes and for long exposure times, as is the unintensified fiber-optic-output image converter with contact film recording.

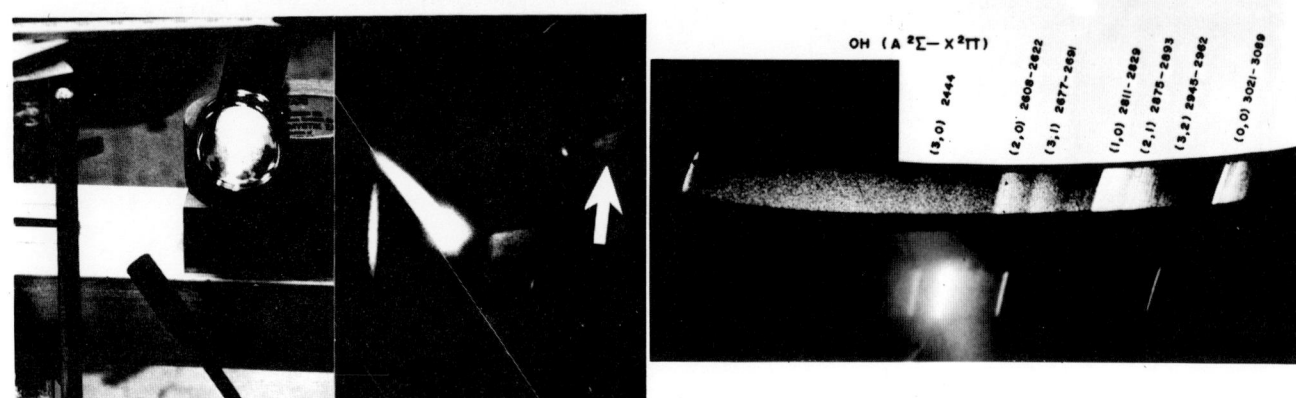

Fig. 5. Comparison of a visible-light photograph and a photograph made with the image converter of Figs. 2 and 7 using Cs_2Te photocathode and 2750 A interference filter. Note that the propane torch flame is almost invisible vs. room light background in the visible photo, whereas the flashlight (arrow) is almost invisible in the UV photo.

Fig. 6. Spectrum of an oxyhydrogen flame made with the Cs_2Te image converter with an object grating but without auxiliary filter. Note strong continuum extending toward shorter wavelengths. Comparison spectrum is of a mercury lamp (strongest line 2537 A).

Fig. 7. Schmidt image converter of the type shown in Fig. 2, (left) image tube with focusing magnet, (right) field-test setup with film transport, front-end filter, and tripod mount.

We have found, in the process of preparing a number of these devices with both Cs_2Te and Cs_3Sb photocathodes, that the former photocathodes are considerably more stable in the presence of small leaks and outgassing than are the latter. Therefore, it should be easier to use the middle-UV photocathodes with electrographic recording than visible-sensitive photocathodes. To our knowledge, there have been no previous attempts to use Cs_2Te photocathodes in electrographic cameras. Therefore, we are presently initiating preliminary investigations into the use of an electrographic Schmidt camera with Cs_2Te photocathode. Since the photocathode cannot be exposed even to dry air, it will be necessary to use an airlock arrangement for emplacing and retrieving film, as in the Kron camera[12]. Ordinarily, the photocathode must also be protected from emulsion outgassing by either cooling the emulsion to liquid nitrogen temperature, or by the use of an electron-permeable barrier membrane in contact with the emulsion. We plan to use the barrier membrane approach, but it is our hope that it will be possible to use membranes and vacuum-system components bakeable at lower temperatures (about $200°C$) than is necessary with visible-sensitive photocathodes ($300-400°C$).

Astronomical Applications

There are several astronomical uses of our middle-UV cameras in ground based work (particularly from high-altitude observing sites), or for use in high-flying airplanes or balloons. The Cs_2Te cathode has useable sensitivity out to 3200 A, but has negligible response to the sky background (in particular the OI 5577 A and the Hg 4358 and 5461 A lines). The room-temperature thermionic emission from Cs_2Te is much lower than that from most visible-sensitive photocathodes. The cameras could be used to study hydroxyl emission in the O-O band near 3090 A, which is one of the strongest cometary emissions, and which may be present in meteor trails. The moderate resolution and low f-number make the device particularly suited to hydroxyl coma measurements, while the wide field of view and low f-number are an advantage for meteor observations.

The primary application to astronomy of these devices, however, is for space observations above the ozone layer. We hope to extend our program of ultraviolet imagery and spectrography into the 2000-3000 A wavelength range, in particular so as to provide a capability for three-color ultraviolet sky surveys to very faint limiting magnitudes. The scientific rationale for this is seen by inspecting the ultraviolet extinction curve for interstellar dust which has been determined from observations with the second Orbiting Astronomical Observatory[13]. A prominent maximum in the extinction curve appears near 2200 A (Fig. 8). Longward of this peak, and also shortward of about 1600 A, extinction increases toward shorter wavelengths (interstellar reddening). However, between 2200 A and 1600 A, the extinction <u>decreases</u> toward shorter wavelengths (interstellar "blueing").

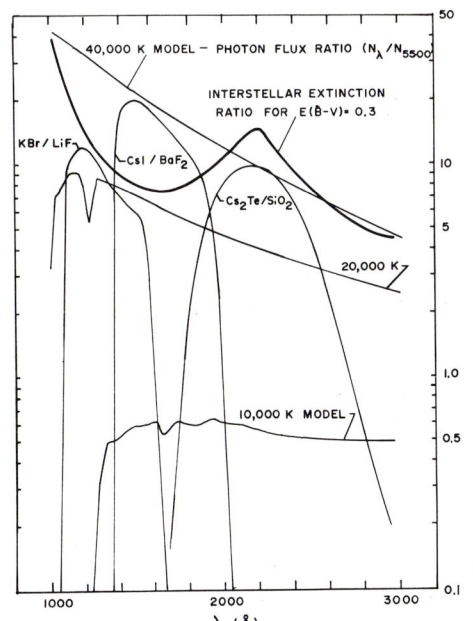

Fig. 8. Curves comparing (1) photon flux vs. wavelength of early-type stellar atmosphere models[14], (2) extinction ratio (I_0/I) vs. wavelength for $E(B-V) = 0.3$[13], and quantum efficiency vs. wavelength for a typical three-color system of opaque-photocathode Schmidt image converters or electrographic cameras.

With alkali halide photocathodes, it is possible to obtain two-color imagery in the 1050-1600 and 1350-1900 A ranges, roughly corresponding to the steep short-wavelength portion and the intermediate minimum portion of the extinction curve, respectively. However, with such a two-color system, it is not possible to separate the effects of stellar temperature from the effects of interstellar extinction (unless the interstellar reddening and/or temperature are known from observations longward of 3000 A). Provision of a third color, centered approximately on the interstellar extinction peak, would provide an independent method of separating temperature and extinction effects in the ultraviolet.

These devices would also be useful for studies of diffuse nebulosities in the 2000-3000 A wavelength range. Reflection nebulosities should show a prominent dip in their spectra near 2200 A, which corresponds to the absorption peak due to interstellar dust (which should also be a minimum in the dust reflectance curve).

These cameras also can be used with an objective grating for spectrographic sky surveys, or in echelle spectrographs for high-dispersion work on large space telescopes. The latter application could also be fulfilled in the near UV-near IR range on ground-based telescopes with the appropriate photocathode materials.

We thank Dr. T. A. Chubb for many useful discussions. This work is supported by the Naval Electronic Systems Command.

References

1. Carruthers, G. R., "Magnetically Focused Electronographic Image Converters for Space Astronomy Applications", Applied Optics 8, 633, 1969.

2. Carruthers, G. R. and Opal, C. B., "Electrographic Cameras for Space Astronomy", in Instrumentation in Astronomy, Vol. 28 of Proceedings (Society of Photo-Optical Instrumentation Engineers), p. 203, 1972.

3. Carruthers, G. R., "Electrographic Cameras for the Vacuum Ultraviolet", in Electrography and Astronomical Applications (University of Texas, Austin), p. 93, 1974.

4. Henry, R. C. and Carruthers, G. R., "Far-Ultraviolet Photography of Orion: Interstellar Dust", Science 170, 527, 1970.

5. Weber, S. V., Henry, R. C., and Carruthers, G. R., "Far-Ultraviolet Interstellar Absorption in Orion and Monoceros", Astrophys. Journal 166, 543, 1971.

6. Opal, C. B., Carruthers, G. R., Prinz, D. K., and Meier, R. R., "Comet Kohoutek: Ultraviolet Images and Spectrograms", Science 185, 702, 1974.

7. Carruthers, G. R., "Apollo 16 Far-Ultraviolet Camera/Spectrograph: Instrument and Operations", Applied Optics 12, 2501, 1973.

8. Carruthers, G. R., Opal, C. B., Page, T. L., Meier, R. R., and Prinz, D. K., "Lyman-α Imagery of Comet Kohoutek", Icarus 23, 526, 1974.

9. Carruthers, G. R., "Microchannel Intensified Electrographic Cameras", presented at Topical Meeting on Imaging in Astronomy, Harvard University, June 1975.

10. Carruthers, G. R., Kervitsky, J., and Opal, C. B., "Some Applications of Microchannel Plates to Electronic Imaging Devices", presented at Sixth Symposium on Photo-Electronic Imaging Devices, Imperial College, London, September 1974.

11. Fisher, G. B., Spicer, W. E., McKernan, P. C., Pereskok, V. F., and Wanner, S. J. "A Standard for Ultraviolet Radiation", Applied Optics 12, 799, 1973.

12. Kron, G. E., Ables, H. D., and Hewitt, A. V., "A Technical Description of the Construction, Function, and Application of the U. S. Navy Electronic Camera", in Advances in Electronics and Electron Physics, Vol. 28A, (Academic Press), p. 1, 1969.

13. Bless, R. C. and Savage, B. D., "Ultraviolet Photometry from the Orbiting Astronomical Observatory. II. Interstellar Extinction", Astrophys. Journal 171, 293, 1972.

14. Kurucz, R. L., Peytremann, E., and Avrett, E. H.,"Blanketed Model Atmospheres for Early-Type Stars", (Smithsonian Institution) 1974.

Session 4
APPLICATIONS OF LOW LIGHT LEVEL DEVICES IN MEDICINE

Session Chairman
Sol Nudelman
Arizona Medical Center, University of Arizona

HIGH-RESOLUTION LOW-LIGHT-LEVEL VIDEO SYSTEMS FOR DIAGNOSTIC RADIOLOGY

H. Roehrig,* M. Frost,† R. Baker,* S. Nudelman,** and P. Capp†
University of Arizona
Tucson, Arizona 85721

Abstract

A high-resolution low-light-level (LLL) video system for use in diagnostic radiology is described. Experimental data are presented that clearly reveal the need for a single-stage LLL intensifier.

Introduction

A general study is under way at the University of Arizona to explore the usefulness of high-resolution video systems for diagnostic radiology. Various factors are being examined to determine the extent to which video can be successful in replacing film as the recording photosensor. These include resolution, minimal discernible contrast, dose reduction, psychophysical limitations, and cost. This paper deals with measurements on our high-resolution vidicon camera system coupled to an x-ray intensifier, and clearly demonstrates that (1) a low-light-level (LLL) condition exists for the high-resolution camera tube, and (2) significant dose reduction is achieved when an LLL intensifier is used with the camera tube. (The camera is an SRL model 352, which uses the Westinghouse 2-in. vidicon WX-5140.)

The high-resolution requirement for diagnostic radiology is easily understood when one considers the TV line number that prevails for an x-ray intensifier and an x-ray screen. For example, the Siemens x-ray intensifier RBV17H has a 7-in. input diameter, a 1-in. output diameter, and a rating of 4.4 lp/mm resolution referred to the input surface. This corresponds to approximately 1500 TV lines, clearly beyond the limits of standard broadcast cameras. Another example is the 14-in. × 17-in. screen used for chest radiographs. The 14-in. dimension encompasses 3500 TV lines for a conservative 5 lp/mm resolution, and 7000 TV lines for a high-resolution screen of 10 lp/mm. Fortunately, such line numbers do not appear impracticable since high-resolution cameras designed by Otto Schade [1] and Kurt Schlesinger [2] have already demonstrated their capability for such tasks.

The basic laboratory facility assembled to date at the University of Arizona for all electronic x-ray imaging comprises an Eimac Microfocal spot x-ray source (providing focal spots of about 50 μm and 0.6 mm), a Siemens x-ray source (providing focal spots of either 0.3 mm or 1.3 mm), and 170-mm Siemens RBV17H x-ray intensifier with about 1500 TV lines resolution, a 40-mm single-stage low-light-level intensifier with an excess of 3000 TV lines resolution and a 35-mm Westinghouse vidicon with an excess of 3000 TV lines resolution per maximum sensor diameter. The image can be viewed on an SRL display with more than 2000 TV lines capability or it can be recorded on 16-mm film with the aid of an electron beam recorder that at present can resolve 1800 TV lines per scan at 50% MTF. These facilities permit a wide range of applications and are being used particularly for magnification radiography.

This paper is concerned specifically with the role of the LLL intensifier. Data obtained will show that our x-ray intensifier-vidicon system needs an LLL intensifier to function properly, and that further increases in resolution requiring larger diameter camera tubes quickly lead to an increasing need for the LLL intensifier.

Technical Discussion

Figure 1 shows what is commonly called a fluoroscopic system in which the x-ray image can be observed instantaneously. X-ray imaging is basically a shadow projection. If the x-ray source were a point source, the transition from shadow to light would be ideally sharp. However, practical x-ray sources cannot be considered point sources. Consequently, the edges of the shadow are fuzzy, resulting from the prevalence of penumbra.

The fuzziness is often referred to as "unsharpness" and is related to the focal spot as indicated schematically in Fig. 1. The penumbra is a function of the imaging geometry. If the object is in the plane of the x-ray intensifier, the penumbra is negligible. However, most objects of interest to the physician are inside the human body or have a finite thickness and cannot be placed in the entrance plane of the x-ray intensifier. This causes a geometric magnification and consequently a sizeable penumbra. Therefore, with a 1.3-mm focal spot and a 75-in. focal spot/object distance (FOD) the x-ray image of a chest is limited to about 5 lp/mm.[3] Imaging for finer detail requires a smaller tube spot size. With the use of the 50-μm Eimac tube to image excised lungs, details as small as 50 μm in the object plane have been successfully visualized, with a geometrical magnification of 10!

---

* Optical Sciences Center
** Department of Radiology and the Optical Sciences Center
† Department of Radiology

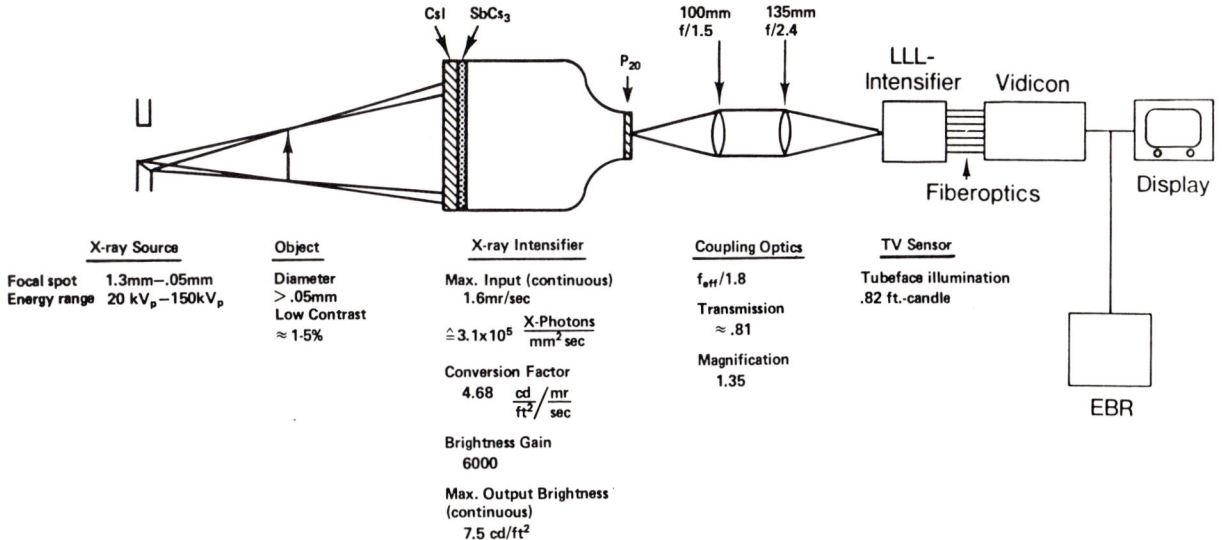

Fig. 1. Schematic and operational characteristics of high-resolution fluoroscopy system.

Next in the chain of components is the x-ray intensifier. It converts the incident x-ray flux into a flux of visible photons. The photosensor is a sandwich structure comprising a fluorescent x-ray sensor such as CsI (which absorbs 60% of the x-ray flux), on which is deposited a conventional $SbCs_3$ photocathode. They convert the fluorescent light into photoelectrons that are then accelerated to the output phosphor. This output phosphor is typically P20 or equivalent, and provides a visible image of the x-ray shadow.

The Siemens tube demagnifies the image from the 170-mm input to the 25-mm output. This demagnification results in a gain that is necessary in order to have a sufficiently bright image at the output, and is called "minification" gain. It is determined by the square of the ratio of the input diameter to the output diameter. For the RBV17H this ratio is 6.8:1; the brightness gain is then about 46.

The x-ray flux or entrance dose rate most typically used in fluoroscopy is about 1 mR/sec in the entrance plane of the x-ray intensifier. Our intensifier has a maximum entrance dose rate of about 1.6 mR/sec for continuous operation. This maximum value is specified to prevent possible permanent image burn-in in the x-ray intensifier output phosphor. The brightness of the x-ray intensifier output phosphor is, in general, a linear function of the x-ray flux at the input. This function is characterized by the conversion factor, which has the units $cd \cdot m^{-2}/mR \cdot sec^{-1}$. For the Siemens x-ray intensifier RBV17H the conversion factor has been measured to be 50 $cd \cdot m^{-2}/mR \cdot sec^{-1}$ or 4.68 $cd \cdot ft^{-2}/mR \cdot sec^{-1}$. Now one can understand the reason for the minification. Using the maximum input dose rate for the x-ray intensifier of 1.6 mR/sec, one obtains a brightness of about 80 cd/m^2. This is not a very large brightness, and without the minification gain, its value would be only about 1.7 cd/m^2.

There is also an average brightness gain specified--this is the ratio of the output phosphor brightness of the fluorescent CsI layer. For the above x-ray intensifier that gain is about 6000.[4]

It is interesting to note that for a typical x-ray spectrum of 80 kVp and an aluminum filtration of 2 mm, an irradiation of 1 mR/sec corresponds to about 2×10^5 x-ray photons/mm^2sec.[5] This is much smaller than the photon fluxes encountered in low-light-level imaging and therefore, one frequently finds quantum limitations on image quality.

The spatial resolution is also an important performance characteristic. It is basically determined by the CsI layer with contributions from the electron optics and the output phosphor. Typical values are on the order of 4.8 lp/mm with improvements up to 8 lp/mm expected in the near future.[4,6]

While most radiologists think of these numbers as representing high resolution, scientists and engineers working in LLL applications, who are used to 20, 30, or more lp/mm, are likely to disagree. However, if one considers the resolution in TV lines per sensor diameter, one obtains a rather impressive 1500 TV lines. Ordinary vidicons cannot handle this kind of resolution, and this is where the need for the high-resolution video system prevails.

It is also worthwhile to dwell briefly on the coupling optics. In typical fluoroscopic systems the visible image appearing at the output phosphor is coupled through a tandem lens system to the television sensor. The first lens is called the collimating lens because the output phosphor is located in its focal plane. The second lens is called the imaging lens; it provides an image of the output phosphor in its focal plane. This arrangement is preferred over a single lens because it collects a maximum of the light emitted by the output phosphor, more than 10 times as much as a single lens. In addition, the parallel beam between the two lenses permits easy insertion of beam splitters. Assuming that the collecting aperture of the imaging lens is larger than that of the collimating lens, the illumination in the image plane of the imaging lens is given by

$$E = \pi \times L \times \left(\frac{f_c}{f_i}\right)^2 \times \frac{\tau_c \times \tau_i}{4(f\#)^2}, \qquad (1)$$

where L = brightness of the phosphor, f_i = focal length of the imaging lens, f_c = focal length of the collimating lens, τ_i = transmission of the imaging lens, τ_c = transmission of the collimating lens, and $f\#$ is the f/No. or the relative aperture of the collimating lens.

The relative aperture of the collimating lens is defined by

$$f\# = \frac{f_c}{D_c}, \qquad (2)$$

which is the ratio of the focal length f_c to the diameter D_c of the collecting aperture of the collimating lens.

For the case when the diameter D_i of the collecting aperture of the imaging lens is smaller than that of the collimating lens D_c, an effective $f\#_{eff}$ has to be defined with

$$f\#_{eff} = \frac{f_c}{D_i}, \qquad (3)$$

and the illumination in the focal plane of the imaging lens is given by Eq. (1) with $f\#_{eff}$ replacing $f\#$:

$$E = \pi \times L \times \left(\frac{f_c}{f_i}\right)^2 \times \frac{\tau_c \times \tau_i}{4(f\#_{eff})^2}. \qquad (4)$$

Most commonly lenses used have a low f/No. (like f/0.65). However, our system uses a collimating lens with f/1.5, 100-mm focal length, and an imaging lens with f/2.4, 135-mm focal length. This combination results in an overall $f\#_{eff}$ of about f/1.8. The choice of lenses was dictated in large part to accommodate a magnification requirement, since the output from the 25-mm phosphor surface of the x-ray intensifier had to be imaged onto the 35-mm surface of the vidicon photosensor. The resultant f/No. was due to the diameter of the imaging lens aperture being smaller than that of the collimating lens.

Originally it was hoped to use just our vidicon as the video sensor in the fluoroscopic chain. However, it was learned quickly that the tube would be light starved, a need for intensification existed, and the amount of intensification would be on the order of 10 to 20.

Examination of the light transfer data of the optics and the maximum brightness of the x-ray intensifier output reveals that the photosensor irradiance of the camera tube is about 0.82 ft candle. Considering the spectral distribution of the P-20 phosphor, the video signal of the 2-in. vidicon operated at 30 frames/sec will not be much higher than about 130 mV maximum. Unfortunately, this is a very low value and provides a very poor signal-to-noise ratio. It was therefore decided to employ a single-stage LLL intensifier.

Experimental Results

In the following, some experiments will be described that were designed to demonstrate the importance of the LLL intensifier. In general, these experiments compare the performance of the fluoroscopic system with and without the LLL intensifier.

First, an aluminum step wedge was used that provides an x-ray gray scale similar to the gray scales used in the visible spectral region. It has 14 steps, each 1/8-in. thick and provides a range of x-ray irradiances of about two orders of magnitude.

This step wedge was used to demonstrate the different video signals obtained for the intensified vidicon compared to the vidicon alone. Figure 2 shows the video line across the step wedge for the case of the vidicon alone. Here the maximum signal is about 300 mV. Figure 3 shows the video signal for the case of the intensified vidicon. The signal is more than twice as high; to appreciate this it is important to point out that the x-ray dose rate was the same and that the f/No. setting for the intensified vidicon was f/4.5 compared to f/1.8 for the vidicon alone. In both cases, one could see up to 12 steps of the aluminum step wedge.

Another test was concerned with the minimum detectable contrast. It was pointed out at the beginning that the minimum resolvable contrast is an important performance criterion in diagnostic medicine. Contrasts are in general very small and often have to be increased with injection of contrast media. A common absorbing material used in evaluating x-ray imaging systems is aluminum. The smaller the minimum detectable aluminum thickness, the better is the contrast performance of the system. One of the experiments performed was to check the relative change of the video signal as a function of the thickness of aluminum placed in the x-ray beam.

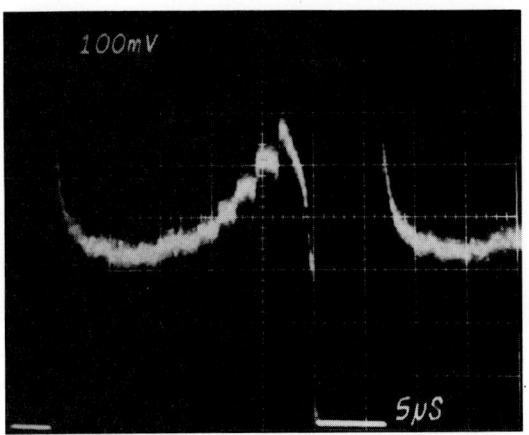

Fig. 2. Video signal of vidicon in response of fluoroscopic system to aluminum step wedge.

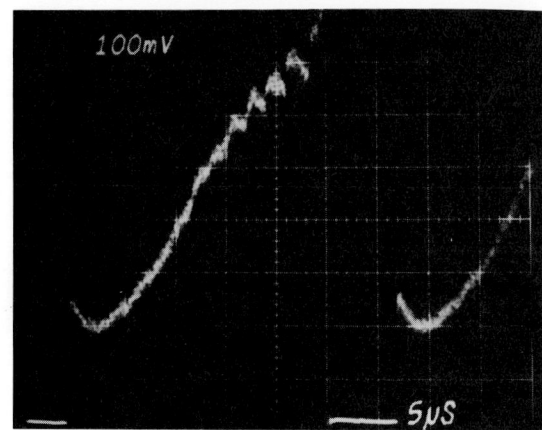

Fig. 3. Video signal of intensified vidicon in response of fluoroscopic system to aluminum step wedge.

The results are shown in Fig. 4, where in both cases the relative change of the video signal is almost linear with the absorber thickness for small aluminum thicknesses. The vidicon alone has a higher slope, which probably comes from the fact that its signal transfer characteristic bends toward a higher gamma at the lower illumination levels. This would mean that the detectable contrast is better for the case of the vidicon than for the vidicon-LLL intensifier combination. However, one has to consider also the accuracy in measuring the signal. The video signal was measured using a line selector A-scope. The signals measured for the vidicon alone were estimated to be accurate to about 12.6%; those for the intensified vidicon, however, were accurate to about 4.8%. This becomes obvious by inspecting the video signals displayed in Figs. 2 and 3. These estimates on accuracy and slope of the curves in Fig. 4 lead to the following values of minimum detectable aluminum thickness as shown in Table 1.

Table 1. Values of minimum detectable aluminum thickness

	Slope	Accuracy	Minimum Al thickness	Dose rate
Vidicon alone	8%/mm	12.8%	$t_{AL} \simeq 1.6$ mm	4.2 mR/sec
Intensified vidicon	5%/mm	4.8%	$t_{AL} \simeq 1$ mm	1.04 mR/sec

Thus, the intensified vidicon can be considered the better device with respect to contrast performance. More important, however, is that this contrast detectability is achieved at a considerably lower dose rate: 1.04 mR/sec compared to 4.2 mR/sec for the vidicon alone.

Figure 5 gives an overview of the x-ray imaging situation. It is a plot of the video signal obtained as a function of the x-ray intensifier dose rate. Indicated are the maximum permissible dose rates at the x-ray intensifier input. Clearly, the intensified vidicon provides images at much lower dose rates than does the vidicon alone. The curve for the intensified vidicon was obtained for an f/No. setting of f/1.8, which permits using a maximum gain of about 70. Obviously at the higher dose rates this gain is

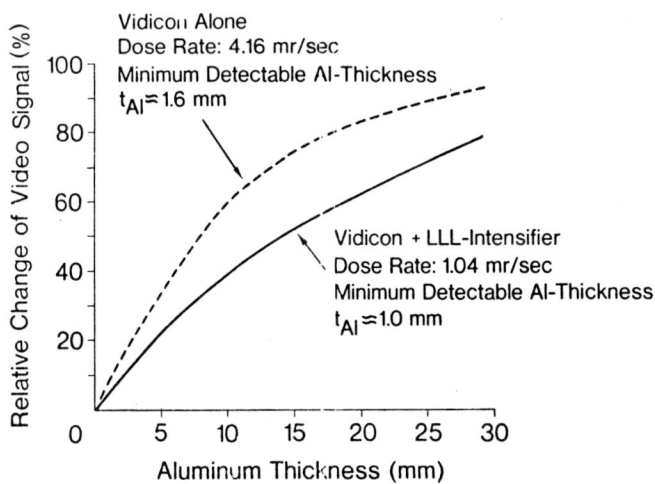

Fig. 4. Relative change of video signal as a function of aluminum thickness placed in the x-ray beam for vidicon and intensified vidicon.

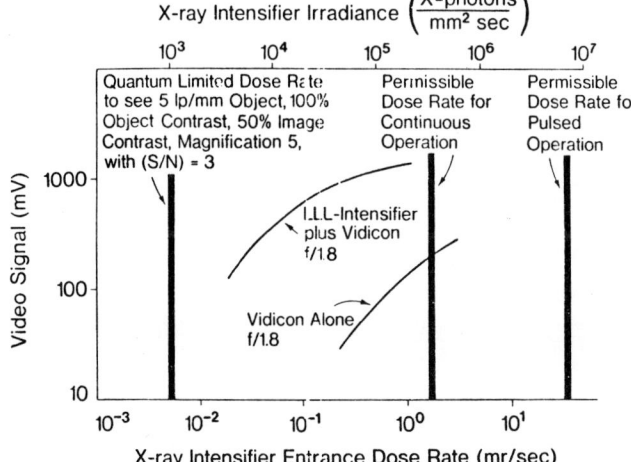

Fig. 5. Video signal as a function of the x-ray intensifier entrance dose rate for vidicon alone and for intensified vidicon.

too high and the curve tends toward saturation. In this case, the gain has to be adjusted by stopping down the lens. The graph also indicates the lower limit of useful dose rates. As is well known, this lower limit is determined by the quantum limitations in the x-ray flux.(7,8,9) Chosen here are the limitations for a 0.1-mm-diameter object, corresponding to a spatial frequency of 5 lp/mm, with 100% object contrast, 50% image contrast, magnification of 5, and a threshold signal-to-noise ratio of $(S/N)_{th} = 3$, which is intended for illustration and to be only a guideline.

Figure 6 is a plot of the signal transfer curves for the vidicon and intensified vidicon. It is interesting to note that the curves of Fig. 5 are almost identical to those of Fig. 6.

Figure 7 is a plot of the resolution capabilities of the most important components of the system, with data compiled from curves found in the literature.(4,10,11) The resolution is either in TV lines per diameter or in lp/mm referred to the x-ray intensifier input. The x-ray intensifier has about 4.5 lp/mm at 5% MTF. The vidicon has about 9.8 lp/mm at 5% MTF, which translates to about 3500 TV lines per 35-mm diameter.

Unfortunately, an MTF for the LLL intensifier was not available, but indications are that it is equivalent to that of the vidicon. The coupling optics also have an appreciable influence on the resolution.

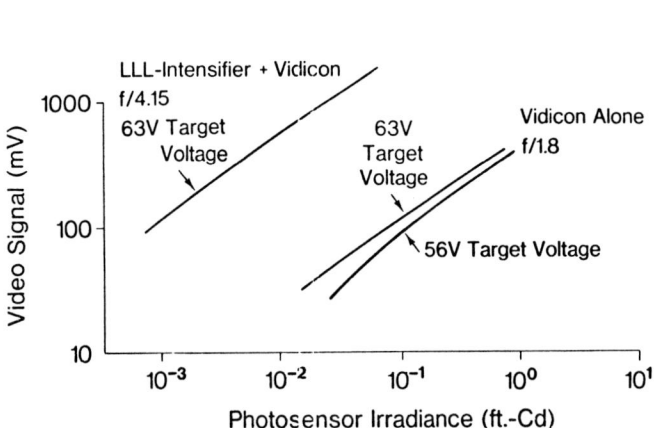

Fig. 6. Signal transfer curves of vidicon and intensified vidicon.

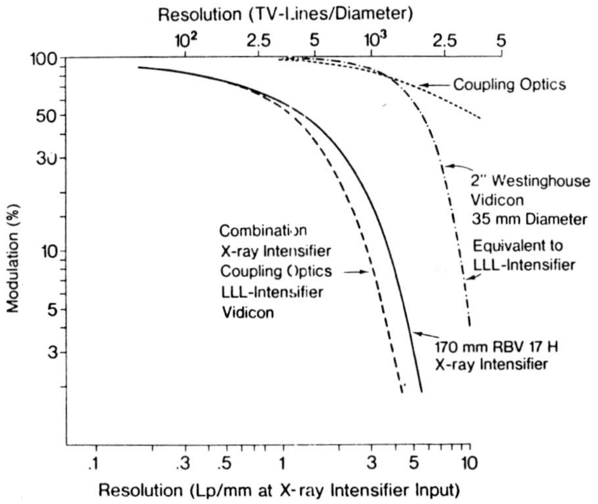

Fig. 7. MTF of the components of the fluoroscopic system and that of their combination.

Using these data the MTF of the overall system was calculated. It is shown by the dotted line in Fig. 7. One would then expect a resolution of 3.5 lp/mm at 5% MTF or a total of about 1100 TV lines per 170 mm of the x-ray intensifier input. Unfortunately, this specification has not yet been reached. While all the components perform according to specifications, the system does not. To date, values up to 2.5 lp/mm have been measured with variations down to about 1.8 lp/mm. However, this does not seem surprising if one considers the fact that the system is composed of parts that were not designed for each other, and that severe alignment and adjustment problems still prevail.

Acknowledgment

This work was funded by the National Heart and Lung Institute, SCOR grant HL14136. These exploratory studies would not have been possible without the 40-mm single-stage intensifier provided by the Night Vision Laboratory through the efforts of Mr. C. Freeman. We are thankful for that assistance, and note that it clearly helped establish the need for low-light-level intensification to achieve high-resolution diagnostic radiology.

References

1. Schade, O., "Theory of operation and performance of high resolution return beam vidicon cameras-- a comparison with high resolution photography," in Photoelectronic Imaging Devices, Vol. 2, L. Biberman and S. Nudelman, Eds., Plenum Press, New York, 1971, pp. 401-437.
2. Saldi, I. T. and Schlesinger, K., "The F.P.S. vidicon," Optical Spectra, Feb. 1970.
3. Roehrig, H., Mockbee, B., Ovitt, T. W., and Freundlich, I. W., "Film-screen combinations in chest radiography," paper presented at the First Image Receptor Conference, Film Screen Combinations, Washington, D.C., Nov. 13-15, 1975.
4. Sirecon RBV17H, Technical Data Sheet MT 501 1178.444, Siemens Aktiengesellschaft, Erlangen, W. Germany.

5. Waggener, R. G., Levy, L. B., Rogers, L. F., and Fanca, P., "Measured x-ray spectra from 25 to 110 kVp for a typical diagnostic unit," Radiology 105:169, 1972.
6. Bates, C. W. and Sparks, S. D., "MTF of a 210-70 mm fiber-optic output image intensifier tube," Applied Optics 14:1484, 1975.
7. Rose, A., "The sensitivity performance of the eye on an absolute scale," Journal of the Optical Society of America 38:196, 1948.
8. Coltman, G. W., "Scintillation limitations to resolving power in imaging devices," Journal of the Optical Society of America 44:234, 1954.
9. Siedband, M. P., "Choosing and setting up TV systems for fluoroscopy," Proceedings of the Society of Photo-optical Instrumentation Engineers, Vol. 35, p. 159, 1972.
10. Technical Data Sheet, Optische Werke, G. Rodenstock, Munchen, W. Germany.
11. Technical Data Sheet TD 86 866, October 1969, Westinghouse Electric Co., Electronic Tube Division, Elmira, NY 14902.

ISOCON IMAGING FOR X-RAY DIAGNOSTICS

Donald Sashin, Ph.D., David Gur, B.Sc., Clive W. Morris, M.B.B.S., John L. Ricci, M.S.

Department of Radiology
School of Medicine
University of Pittsburgh
Pittsburgh, Pennsylvania 15261

Abstract

The disadvantages of full size film and film screen techniques have stimulated research during recent years for an alternate approach. For our technique, we have been developing a low light level television system to eliminate many of the disadvantages thereby yielding the potential for improved diagnosis of early pathology at reduced radiation, cost, procedure time and storage and retrieval expense. This isocon imaging system is being developed for early breast cancer, lung cancer and pneumoconiosis detection.

Introduction

In order to improve diagnostic radiology there has been continual research in alternative methods for recording, displaying and storing the diagnostic radiographic image. Some of the disadvantages of the present film cassette and non-screen film methods are the cost, the radiation exposure, the difficulty of storage and the loss of low contrast detail due to limitations in these techniques. For our technique, we are developing a low light level television imaging system for improved diagnosis and reduction of radiation exposure through the use of an image isocon television camera optically focused at a high detail fluoroscopic screen. Our approach has been concerned with the development of a method which does not use any image intensifier thereby eliminating the degradation in contrast resulting from its use. Furthermore, the large x-ray intensifier has the additional disadvantages of a round format and relatively high cost.

In order to develop a system having a high quantum effeciency with acceptable noise, image latitude, contrast and frequency response, we are conducting a comprehensive study to investigate the relevant parameters of the apparatus. With this information, component and design changes are incorporated into the system to evolve a final system which will record and display electronic radiographic images with increased diagnostic information at reduced x-ray radiation levels to the patient and personnel.

The isocon imaging systems are presently being developed in our laboratory for application to early breast cancer detection, as well as pneumoconiosis and lung cancer detection. In all of these systems the basic technique of obtaining the electronic radiographs is similar.

Method and Procedure

The principle of operation of the isocon imaging system is the following (see Figure 1): the imaging sequence is initiated by the closing of a single pole, single throw push button switch that is located in a "mini-box" connected to the apparatus by a 10 foot cable. Once the switch is depressed, the x-ray generator applies a high voltage to the x-ray tube producing a continuous   x-ray beam. As the x-ray are turned on, the motor for the moving slit collimator is started in the forward direction. This narrow slit produces a fan-shaped beam of x-rays that moves across the anatomical portion of the patient to be imaged. Those x-rays that pass through the patient strike a thin x-ray phosphor screen which converts the radiographic image into a visible image that moves across the screen as the slit collimated beam crosses the patient. This image is focused by a very fast f/.7, 60 mm Canon lens onto the photocathode of the isocon television camera tube (see Figure 2).

---

*This work is supported in part by grants from the American Cancer Society, Grant #58M and PHS-FDA-BRH #FD00657, National Science Foundation, Grant ENG 75-14852, and US-PHS-FD-BRH Training Grant RL-00063.

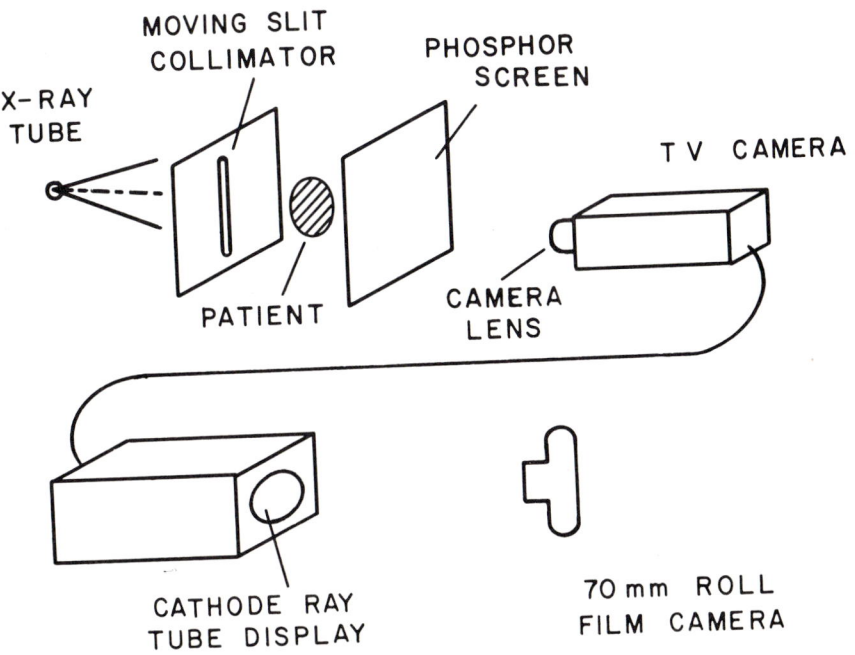

FIGURE 1: Block diagram of isocon imaging apparatus.

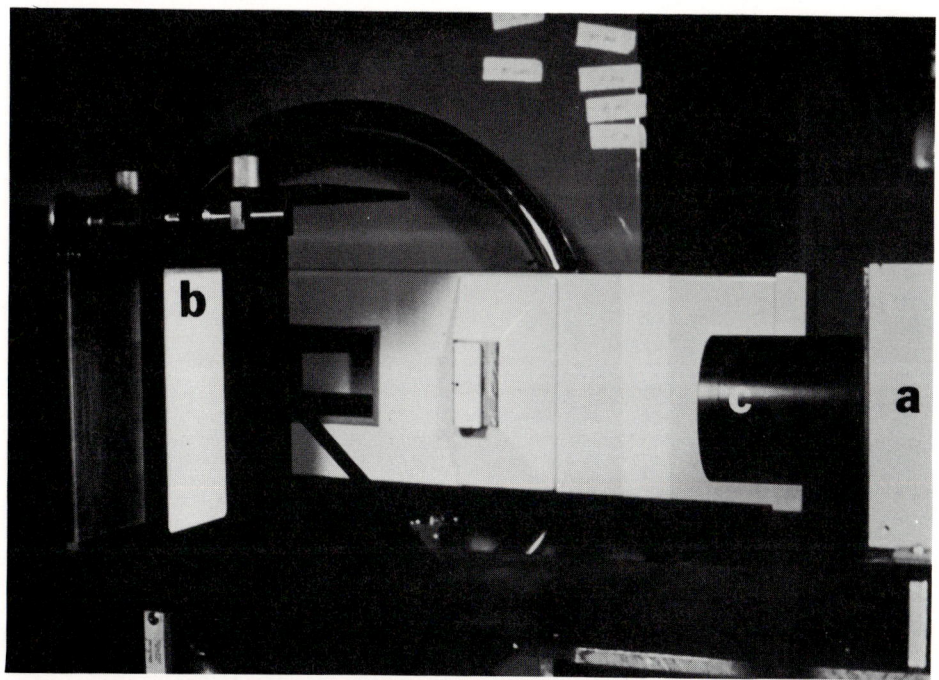

FIGURE 2: This picture has been taken with the cover of the apparatus removed. The isocon camera (a) is focused at the x-ray phosphor screen (b) using a high speed lens (c).

The television camera scans the image at a 30 cycles/second frame rate, with a sequential scan of 525 television lines. The television signal is amplified and displayed on a flat faced 5 inch cathode ray tube. The video signal is selectively blanked synchronously with the motion of the fan-shaped x-ray beam so that only the section of the image formed by the primary beam is actually displayed on the cathode ray tube (see Figure 3). This image is updated on the CRT at 30 frames/second.

FIGURE 3: This picture of the Cathode Ray Tube Display (a) with the 70 mm magazine (b) for photographing the electronic radiographic images.

In this way, the x-ray scatter in the patient, and the scatter and flare of the image in the fluorescent screen, lens and isocon camera tube are electronically removed from the displayed image on the face of the cathode ray tube. In addition, a small astigmatism coil on the neck of the CRT stretches the television lines to completely eliminate their spacing on the screen of the CRT and on the small format film, respectively. The CRT images are photographed with a standard oscilloscope camera onto 70 mm roll film contained inside a film magazine thus facilitating the handling, since the magazine can be easily removed from the photographic camera system and taken into the dark room for processing of exposed film and for reloading of the magazine. In our method, the final image is recorded on the film by photographing 50-60 television frames on the same film. Every image on the film is really an integration of many television images each of which contributes only a few lines of information. The apparatus automatically synchronizes the x-ray generation, the moving slit collimator, the camera scan and the film transport under the control of electronic circuits which are triggered by a single push button switch. Some important features of the system include aperture correction, gamma correction, and low pass filtration circuitry in the television camera control unit to electronically enhance the image and improve the signal-to-noise ratio. In the present system the final image is recorded on film for viewing and analysis. Because the radiographic information is basically in electronic form, it is anticipated that in the future the images would be recorded on either video tape or video disc. Furthermore, with the increased sophistication of computer programming and hardware, there exists the potential for automatic processing and analysis of the images.

During the past few years, we have developed a prototype electronic radiographic system for application to mass screening for mammography (see Figure 4). Many of the operating parameters have been carefully selected as a result of our analysis of the various components of the apparatus under both laboratory and clinical conditions. These studies explored their advantages and limitations in a specific application or electronic radiology for mass screening for early detection of breast cancer (1, 2, 3, 4, 5).

FIGURE 4: This picture demonstrates the compression of the patient's breast against the x-ray phosphor screen (arrow).

Discussion

In order to improve the image quality while simultaneously reducing the radiation dosage, it becomes apparent that the electronic noise inside the television camera and camera tube would have to be reduced. A comprehensive investigation of the target voltage of the isocon camera demonstrated that by increasing the voltage from approximately 3 to 5 volts the signal-to-noise at low light levels would increase by about two times. We were also able to further increase the signal-to-noise by redesigning some sections of the preamplifiers in the isocon camera and by carefully matching the band-pass of the television camera to the information in the electronic radiographic image. In the analysis of the television line rate and the number of scan lines per image, it was demonstrated that 525 lines and 30 frames/second appeared to yield the optimum signal-to-noise for our images. Of course the television scan line structure is removed from electronic radiographs by the spot stretch coil mounted on the neck of cathode ray tube. The prototype electronic mammographic apparatus has been evaluated in a series of small clinical studies to identify advantages and limitations of the apparatus for mass screening for breast cancer. At the present time some of the advantages of this new method are: (a) its low radiation exposure, (b) low cost of operation, (c) reduced film and film processing costs which are becoming a serious problem as the price of silver continues to rise, (d) reduced film storage costs (e) short examination time, (f) reduction in the detection of scattered x-rays and scattered light onto the final image through the use of the electronic moving slit technique, (g) electronic image enhancement, aperture correction, and low pass filtration, (h) the potential for computer image enhancement and eventually automatic analysis.

In the clinical studies, the electronic mammographic images had adequate clarity to demonstrate cancers, gross calcification, tumors, skin thickening and trabecular structures. Calcifications have been demonstrated, but at present very fine calcifications are not adequately visualized. The radiation exposure ratio for the electronic mammogram compared to the Lo-Dose film screen method was 0.17 ± 0.06. From these studies we have concluded that the detail of electronic images is presently being limited by the noise in the system. In preliminary studies being conducted in our laboratory, it appears that this noise can be further reduced which would improve the image clarity while significantly reducing the radiation dosage below 100 mR per image.

References

1. D. Sashin, C. W. Morris, "Electronic Mammography for Early Tumor Diagnosis", <u>Radiological and Other Biophysical Methods in Tumor Diagnosis</u>, Year Book Publisher, Inc., pp. 415-421, February, 1975.

2. J. O. Barnes, W. H. Kennedy, D. Sashin, J. L. Ricci, A. Porti, "Imaging Properties of an Electronic Mammographic System", <u>Application of Optical Instrumentation in Medicine III</u>, Vol. 47, pp. 196-201, 1975.

3. J. L. Ricci, J. O. Barnes, D. Sashin, "Contrast and Dose Evaluation in Mammography by Means of Spectral Analysis", <u>Application of Optical Instrumentation in Medicine III</u>, Vol. 47, pp. 183-195, 1975.

4. D. Sashin, C. Morris, J. L. Ricci, J. O. Barnes, "An Evaluation of Low Light Level Television for Breast Cancer Detection", <u>Application of Optical Instrumentation in Medicine IV</u>, Vol. 70, pp. 384-392, 1976.

5. D. Sashin, J. O. Barnes, J. L. Ricci, D. Gur, "Electronic Moving Slit for Scatter Reduction in Diagnostic Radiology", <u>Operational Health Physics</u>, pp. 787-790, February, 1976.

IMAGE INTENSIFIER SCINTILLATION CAMERAS
FOR NUCLEAR MEDICINE APPLICATIONS

Gerd Muehllehner, Ph.D.
Searle Radiographics Inc. and Searle Analytic Inc.
2000 Nuclear Drive, Des Plaines, Illinois 60018

Abstract

Various scintillation cameras using image intensifiers are described and compared in their important imaging characteristics with existing instrumentation. The main advantages of image intensifier cameras are in the areas of positional resolution, count-rate capability and system simplicity. The main problem in the past has been the general lack of good gamma-ray energy discrimination which results in low contrast images in clinical situations. Good energy discrimination is in principle possible with several of the described systems.

Introduction

For many years the Anger scintillation camera has been the standard device to image radionuclide distributions in the field of nuclear medicine and it has remained the instrument of choice so far in spite of many shortcomings and attempts to use alternate approaches. In particular, various configurations using image intensifiers have been proposed and built with the intent to develop a device with superior imaging characteristics.

The important performance parameters and typical values presently achieved with the Anger camera are the following. Positional resolution of the detector - degradation due to collimation will be discussed later - has recently been improved from about 8 mm to 5 or 6 mm FWHM by replacing 19 3-inch photomultipliers by 37 2-inch photomultipliers. The field-of-view is generally 25 cm but the trend is towards larger useful diameters in order to be able to image large organs such as the lungs and in order to be able to take full advantage of the desirable imaging properties of converging collimators for smaller organs.[1] Maximum count-rate capability is important in fast dynamic studies such as some cardiac studies and has resulted in devices with a dead time of approximately 2 microseconds leading to a maximum rate of 200 K cts/sec which, however, is reduced in the presence of scatter. In order to eliminate scattered radiation from the patient energy discrimination is essential to avoid loss of contrast; 16% energy resolution FWHM of the photopeak at an energy of 140 keV still leads to the inclusion of many (38%) unwanted scattered events.

The areas where image intensifiers are most likely to advance the present level of performance are positional resolution and count-rate capability. An additional factor should not be underestimated: An Anger camera with 37 photomultipliers and the associated electronics is a complicated piece of equipment and an image intensifier camera could potentially lead to a greatly simplified device. Image intensifiers have positional resolution which far exceeds that presently achieved in radioisotope cameras, even if the resolution is degraded to 2 mm FWHM by the scintillation crystal, it still represents a significant improvement over present devices. In order to assess the overall system improvement resulting from this improved intrinsic resolution (R_I) the degradation due to the collimator must be taken into account. Figure 1 shows system resolution as a function of distance from the face of the collimator for various values of intrinsic resolution. It can be seen that the improvement from 5 mm (R_I) to 2 mm (R_I) improves the system resolution by only 1 mm or 20% at a distance of 10 cm from the collimator.

In discussing various device configurations emphasis will be placed on the front-end configuration; often the output is integrated directly on film or alternately detected by a television system. Practically all devices discussed here can use either recording method and the basic performance of the device is little affected by the choice of output medium.

Lens Focussing Cameras

Scintillation cameras using a lens to focus the light from the scintillation crystal onto an image intensifier are attractive because of their simplicity of design. Gamma cameras of this type are exemplified by the Aber-gammascope[2] and the one developed at the University of Michigan[3], but others have been active in pursuing this approach as well.[4, 5, 6] As shown schematically in Figure 2, the light from the scintillation crystal is focussed through a large aperture lens onto a relatively small (up to 50 mm) multi-stage image intensifier and the intensified image is either recorded directly onto film or viewed by a television camera.

Ideally each scintillation event occurring in the crystal should result in the emission of one or more photoelectrons from the photocathode of the image intensifier. A 140 keV photoelectric interaction in NaI(Tl) will yield approximately 5000 photons within a broad spectrum centered around 4100 A°. Using a lens system with commercially available lenses having an effective f-number of 0.75 and a minification of 6:1 (22.8 diameter NaI(Tl) crystal, 38 mm diameter photocathode on image intensifier) the group at the University of Michigan have achieved a photoelectron/gamma ratio of 0.15 to 0.3. While the major light loss is due to the limited solid angle of light collection other important factors include the high index

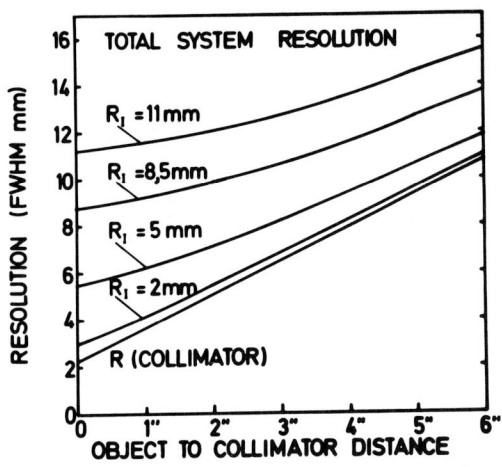

FIG. 1: Total system resolution for various values of intrinsic resolution with a particular collimator.

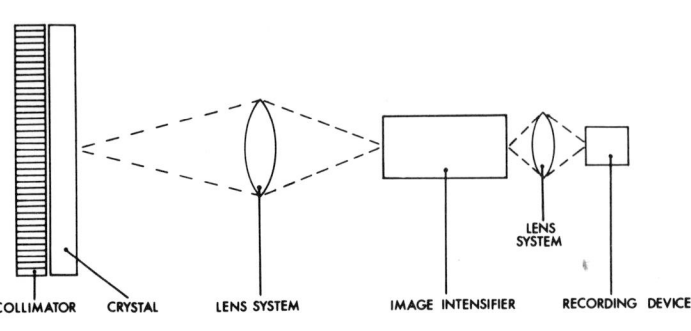

FIG. 2: Typical schematic configuration of lens focussing cameras showing major components only.(2, 3)

of refraction of NaI (1.78) and the quantum efficiency of the photocathode (24%). The group at Aberdeen has reported a photoelectron/gamma ratio of 6 for 140 keV radiation and is using a set of plastic lenses which were specifically designed for the purpose. This is sufficient to record practically all photopeak events occurring in the crystal but is not sufficient to allow this amplified signal to be used for energy discrimination.

While the early experimental versions of this type of camera were used to take images of phantoms of remarkable quality these were usually taken with high concentration of activity and without any scattering material present. In clinical cases the low amounts of activity leading to long exposure times, the presence of surrounding tissue and the often low object contrast showed that the basic concept had to be significantly modified if the device were to be clinically useful. The low contrast in clinical images is mainly due to the lack of energy discrimination to eliminate scattered radiation from the object and due to light being internally reflected in the crystal which on its way to the lateral edge of the crystal escapes from the crystal and leads to a significant background. While the latter can be reduced by careful attention to the state of polish of the crystal surfaces the former requires that some of the light not collected by the lens is collected at the edge of the crystal and detected by several photomultipliers. This signal allows pulse height discrimination to be used to eliminate most scattered radiation and high energy events (cosmic rays) by gating one of the stages of the intensifier. This not only complicates a basically elegant and simple system but also reduces the count-rate capability drastically even if careful attention is given to the selection of the phosphors in the various states of amplification.

The effect of scattered radiation from the patient can also be reduced by placing a set of absorbers[7] in front of the crystal which attenuates low energy scattered radiation to a larger degree than the primary radiation. Figure 3 shows a spectrum of ^{99m}Tc in the presence of scatter material both with and without a graded absorber consisting of 0.9 mm tin and 0.13 mm copper; the copper layer serves the purpose of reducing the intensity of the characteristic X-ray from the tin absorber. This graded absorber attenuates the primary radiation by 33%. One benefit derived from this form of energy discrimination is the freedom to choose a scintillator which is not polished but uses other means such as grooves, dividers, dyes etc. in order to prevent the light from spreading laterally. It also retains the high data rate handling capability of the original concept.

In summary, the lens focussing cameras are inherently simple devices, excelling in positional resolution (2 to 3 mm intrinsic resolution) and having very high count-rate capability. The latter may have to be sacrificed in order to achieve pulse height discrimination.

Cameras with Large Diameter Image Intensifiers

In the previously described approach a major problem is the light loss through the lens system. This can be avoided if a large diameter image intensifier is directly coupled to the scintillation crystal. Now the major problem is the inability of the device to focus the light which originates inside the crystal into a fine dot. Three approaches have been tried to overcome this problem: 1) the scintillator is made thin enough to avoid any significant lateral spreading of the light, 2) the scintillator is made thick but made up of many individual pieces only a few millimeters in diameter separated from each other by opaque material, or 3) the light is allowed to spread into a diffuse spot which is intensified and at a later stage the centroid of the diffuse spot is found and used to represent the location of the interaction in the crystal. Each of these will be discussed in turn.

Thin-crystal Cameras

The best-known development in this category is the camera first developed by M. Ter-Pogossian and later in modified form made commercially available by Picker Corporation.[8,9] In this system a thin (1.5 mm) CsI(Na) crystal is incorporated into a 9 inch diameter intensifier tube, the minified output is focussed onto a SEC tube with an additional single state of intensification at the input (Figure 4). Several variants of this approach have been tried.[10,11,12] One important characteristic of the CsI(Na) crystal is that it can be curved to fit the radius of curvature of the intensifier. The thickness is chosen as a compromise between resolution (intrinsic resolution 2 mm) and sensitivity. The approximate detection efficiency at 140 keV is 40%, dropping rapidly at higher energies; the device is thus largely restricted to use with low energy emitters which, however, is not a major drawback since the majority of all imaging is currently performed with ^{99m}Tc (140 keV). Placing the crystal on the outside of the intensifier[10] leads to serious problems with light spreading laterally in the crystal and the glass entrance window, air coupling the crystal to the glass alleviates some of the lateral spreading but reduces the total amount of light reaching the cathode. Energy discrimination can be achieved to a limited degree by performing pulse height analysis on the video signal[8] and sophisticated schemes of analyzing the video signal to achieve improved energy discrimination have been proposed.[12] This approach has the inherent advantage that two simultaneous events which are recorded in different locations can yet be used and pulse analysis be performed on both. Earlier versions of the device suffered from uniformity problems in both the intensifier and the SEC tube as well as lack of energy discrimination. Later versions have been used in a variety of dynamic imaging situations but high-quality static images have not been reported in the literature, the lack of good energy discrimination probably is the major reason for this. There is, however, no inherent fundamental limitation which prevents good energy discrimination and the approach is still a sound one with many good imaging characteristics.

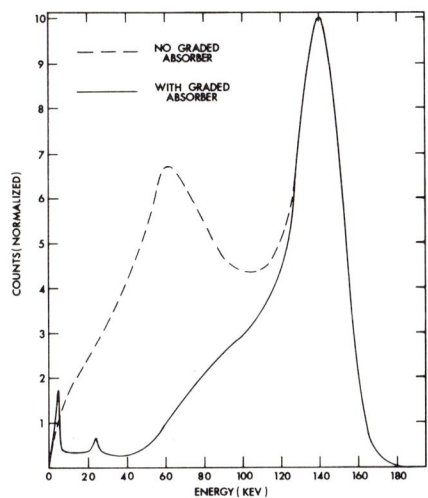

FIG. 3: Influence of graded absorber on energy spectrum of ^{99m}Tc in presence of scattering material as measured with a NaI(Tl) detector.

FIG. 4: Typical schematic configuration of thin crystal cameras.[8,9]

Crystal-mosaic Cameras

To achieve both high gamma-ray stopping power in the scintillator and at the same time avoid lateral dispersion of the light emanating from the point of interaction a crystal mosaic may be placed against the front faceplate of a large image intensifier. The University of Toronto group[13] uses a CsI(Na) mosaic of some 1500 elements which are 4.1 mm square and 2.5 cm long while the camera developed by Harshaw Chemical Co.[14] called Quantascope, uses 2515 individual CsI(Tl) crystals 3.2 mm in diameter and approximately 1.6 cm long. Major components of both cameras are shown in Figure 5. The center-to-center spacing between elements of both devices is 4.76 mm and limits the achievable resolution and image quality while at the same time the large number of individual elements represents a major drawback. Energy discrimination can be achieved by two methods. In the University of Toronto camera the minified image from the first intensifier is coupled through a lens to a gated intensifier. An elliptical plate collects much of the light not collected by the lens system and focusses it onto a photomultiplier whose output is analyzed in a single channel analyzer and is used to turn the gated intensifier on after a desirable event has occurred. The Quantascope uses an interesting switching tube which directs the electrons from the cathode of the second intensifier tube to a fast (P15) phosphor whose output is detected by a photomultiplier and is used after pulse height discrimination to switch the beam to a slow, high-gain phosphor for image formation. This tube could of course be equally well used with other front-end configurations and might work well with a single thin scintillation crystal at the input as in the thin-crystal cameras described above. In all techniques using a gating or switching mechanism in the intensifying chain, proper choice of phosphors is important since the early part of the emitted light is used for energy discrimination while the remainder is used for image formation. These forms of energy discrimination result generally in long deadtimes; while 10 μsec or even less appears feasible, values actually achieved are generally longer (maximum stated data rate for University of Toronto camera is 20,000 cts/sec). In general, the crystal-mosaic cameras appear to offer little advantage over existing instrumentation, since the hoped-for advantages of image intensifier cameras - high spatial resolution, high count-rate capability and simplicity of design - are not inherent characteristics of this approach.

Re-focussing Cameras

In the Anger scintillation camera the light originating in the scintillation crystal is allowed to spread and the signals from a multiplicity of photomultipliers are used to find the centroid of the distributed light in order to display a dot on a CRT corresponding to the location of the centroid. This basic scheme is used in an approach which substitutes an image intensifier for the limited number of discrete detectors (photo- multipliers) shown schematically in Figure 6. Three similar versions were built independently[15, 16, 17] which have the common feature that a thick (12.5 mm) NaI(Tl) scintillation crystal is coupled to a large diameter image intensifier. Either a flat crystal can be coupled through a coarse fiber-optic lightguide or a curved crystal can be coupled directly to the curved entrance window. The main problem encountered in trying to couple the flat crystal through coarse fiber-optics is the reduction in the numerical aperture of the fiber optics which results from the fact that the light originates in a medium with a high index of refraction. Within the last few years Harshaw Chemical Company has developed "Polyscint", a polycrys- talline NaI(Tl) material which can be supplied in the desired curved configuration at a reasonable cost which is said to perform as well as monocrystalline NaI(Tl) and which would eliminate the need for the fiber-optic lightpipe. Potentially the electron beam in the first stage could impinge on an internally locat- ed position-sensitive solid state element instead of a phosphor. While this is technologically feasible, no such image tube is currently available. Instead the light from the first stage - after further intensifica- tion - is detected by an externally located silicon diode detector with the contacts consisting of parallel strips at right angles to each other (Figure 7). The signals from the strips are coupled to four preampli- fiers through resistive or capacitive dividers in an arrangement commonly used with position-sensitive proportional counters. This results in signals which can be used to indicate the position of the centroid

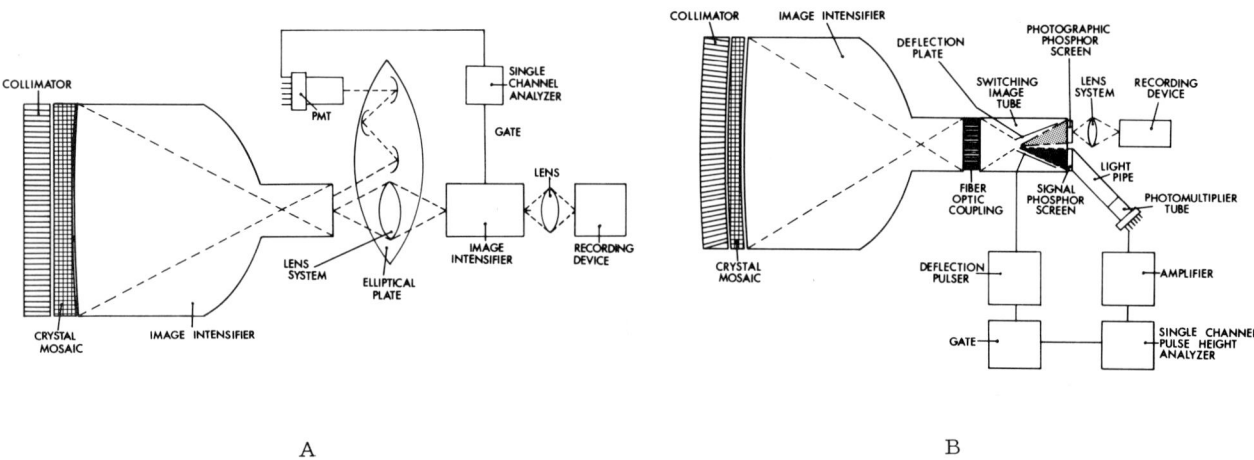

FIG. 5: Crystal mosaic cameras. A) University of Toronto device[13] B) Quantascope[14]

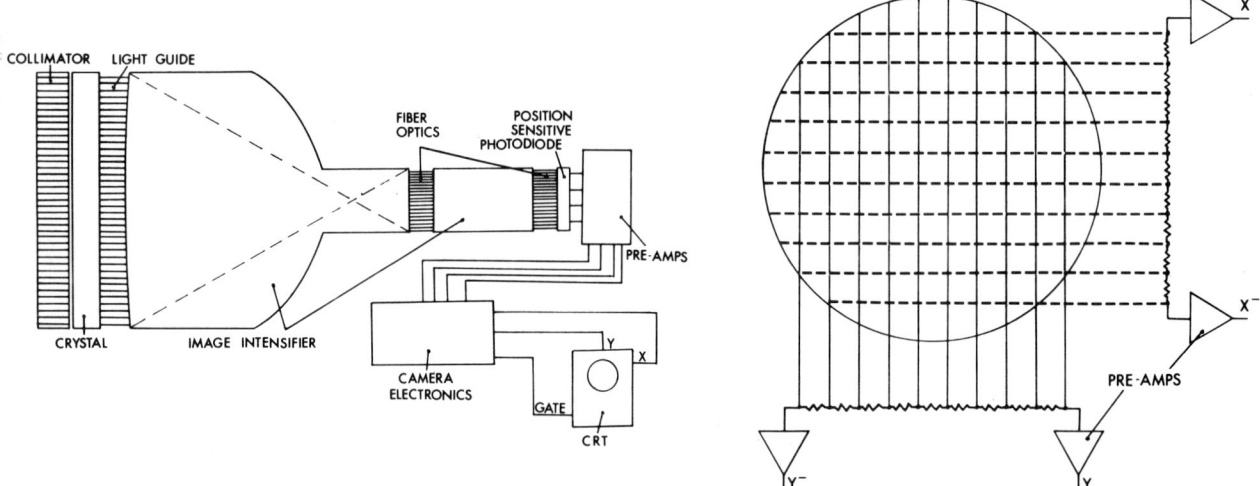

FIG. 6: Typical configuration of re-focussing cameras showing major components.[15] The solid state detector can be replaced by 4 photomul- tipliers.[16, 17]

FIG. 7: Silicon diode detector with orthogonal strips for position readout.

of the light distribution as well as the total amount of light emitted. Thus good energy distribution can be combined with the use of a thick crystal. Instead of the position-sensitive solid state element, the light from the intensifier can be detected by four photomultipliers the output from which are used to find the centroid of the light distribution. A scintillation camera of this type had recently been marketed by Siemens and a similar device has been reported[18] to give high resolution (2.5 mm), high count-rate capability (10^5 cts/sec) and large field of view (35 cm).

Summary

Scintillation cameras using image intensifiers promise several advantages over existing devices, in particular high spatial resolution, high count-rate capability and simplicity of design. Three different versions have been developed to the point where commercial prototypes were clinically evaluated, however, none are now in widespread use. The most important reason for the limited success of image intensifier cameras so far has been the lack of good energy discrimination which leads to low contrast images in clinical situations. However, some of the approaches may well be improved to the point where good energy discrimination can be incorporated.

References

1. Muehllehner G.: Advances in nuclear imaging instrumentation. Proceedings of the First World Congress of Nuclear Medicine (World Federation of Nuclear Medicine and Biology, Tokyo, 1974) 354-361.

2. Mitchell J.G., Mallard J.R., Egerton I.B., et. al.: Towards a fine-resolution image-intensifier gamma-camera, the Aber-gammascope. Medical Radioisotope Scintigraphy, 1972, Vol. I. (IAEA, Vienna, 1973) 157-167.

3. Thomas F.D., Beierwaltes W.H., Knoll G.F., et. al.: A new scintillation camera. Medical Radioisotope Scintigraphy, Vol. I. (IAEA, Vienna, 1969) 43-56.

4. Iio M., Ueda H., Sasaki Y., et. al.: A multistage image intensifier gamma camera and radioisotope angiography. Medical Radioisotope Scintigraphy, Vol. I. (IAEA, Vienna, 1969) 17-29

5. Corry P.M.: Image-intensifier scintillation camera. Bull. Am. Phys. Soc., 12, 675, 1967.

6. Muehllehner G.: unpublished work.

7. Muehllehner G., Jaszczak R.J., Beck R.N.: The reduction of coincidence loss in radionuclide imaging cameras through the use of composite filters. Phys. Med. Biol., 19, 504-510, 1974.

8. Ter-Pogossian M.M.: An image intensifier isotope camera capable of pulse height discrimination. Advances in Medical Physics (Second International Conference on Medical Physics, Boston, 1971) 179-184.

9. Freedman G.S., Goodwin P.N., Johnson P.M., Pierson R.N.: An evaluation of the image-intensifier scintillation camera with some comparisons to the single crystal camera. Radiology, 92, 21-29, 1969.

10. Kellershohn C., Lansiart A., Desgrez A.: Deux nouveaux types de detecteurs pour camera a rayons X ou γ. Medical Radioisotope Scanning, Vol. I. (IAEA, Vienna, 1964) 333-354.

11. Sternglass E.J.: Gamma ray, X-ray image converter utilizing a scintillation camera system. U.S. Patent, No. 3462 601, 1969.

12. Asars J.A., Schneeberger R.J.: Television type nuclear radiation camera system. U.S. Patent, No. 3668396, 1972.

13. Moody N.F., Joy M., Paul W.: An image intensifier gamma-ray camera and its variants. Medical Radioisotope Scintigraphy, 1972, Vol. I (IAEA, Vienna, 1973) 255-267.

14. Harshaw Chemical Co.: Quantascope. Company bulletin.

15. Muehllehner, G.: Radiation imaging apparatus. U.S. Patent, No. 3683185, 1972.

16. Lansiart A., Lequais J., Roux G., et. al.: Detecteur stationaire a gaz et nouveau type de gammascope. Medical Radioisotope Scintigraphy, Vol. I (IAEA, Vienna, 1969) 87-98.

17. Conrad B.: A detector for determining the distribution of radioactive materials. British Patent Specification, No. 1174558, 1968.

18. Driard B., Rozieres G., Verat M.: A large field image intensifier tube for scintillation cameras. IEEE, NS-23, Feb. 1976 (to be published).

METHODS AND NEW APPROACHES TO THE CALCULATION OF
PHYSIOLOGICAL PARAMETERS BY VIDEODENSITOMETRY

Dan Kedem*, Drora Kedem*, D.P. Lindstrom, T.C. Rhea, Jr., J.H. Nelson,
R.R. Price, C.W. Smith, T.P. Graham, Jr., and A.B. Brill
Vanderbilt University Hospital, Nashville, Tn. 37232

Abstract

A complex system featuring a video-camera connected to a video disk, cine (medical motion picture) camera and PDP-9 computer with various input/output facilities has been developed. This system enables the performance of quantitative analysis of various functions recorded in clinical studies. Several studies are described, such as heart chamber volume calculations, left ventricle ejection fraction, blood flow through the lungs and also the possibility of obtaining information about blood flow and constrictions in small cross-section vessels.

Introduction

In the last few years many efforts have been made to perform quantitative rather than qualitative analysis of the data obtained from radiographic images.[1-5] The main reason was the need to extract more information than provided by visual inspection alone. Moreover, the progress made in fast data acquisition, such as recording information gathered in a 3-5 msec video rates (60 fields/sec), made the analysis of fast time-dependent processes a very valuable source of information.

In order to obtain information from such short periods of time, either high dose rates or very sensitive low light level techniques are required. The latter is more likely to be used in medicine, so as to prevent the patients from being exposed to high x-ray dose levels.

The system described in this paper has been adapted for measurements in studies concerning physiological performance and anatomic structure of organs like the heart, lungs or kidneys.

One of the most accurate ways to perform these studies is by the videodensitometry method, provided that care is taken to use this tool in the appropriate manner. Accurate calibration of density is needed only in some studies, where absolute values are required, while calculations of relative parameters or ratios do not require the knowledge of absolute density.

System

The purpose of our system is first to pre-process, store, and digitize video radiographic images, and second to make calculations on them, and to store and display the results in video form too. In our Cardiac Catheterization Laboratory two x-ray tubes are positioned at right angles to each other (Fig. 1). After passing through the patient, each beam is projected onto the front surface of an image intensifier tube. The output phosphor of each tube is then viewed by a motion picture camera and using a partially-silvered mirror, a plumbicon tube. A two-camera video-combiner circuit then creates a split screen image containing both video signals. Another special effects box puts the patient's EKG on the bottom of the combined video picture, and multiplexes sampled values from up to 8 analog transducers on the lateral border of the recorded image.

The video data analysis system consists of a computer, a video disk recorder, and several interfaces which link the disk, the computer, and display devices (Fig. 1). The video disk recorder (Data Disc 310) is the heart of the system, since it provides storage for up to 600 video frames during processing and also provides the necessary timing pulses for digitizing video and converting digital data into video form. In addition to the timing tracks from which video sync is derived, a 5 MHz clock track is recorded on the disk to synchronize AD and DA conversions. Touch tone-encoded keyboard units have been added so that record and playback functions can be controlled from remote locations. The frame number is displayed in a corner of the video image, to aid in positioning the read heads at the desired track.

An interface has been built which allows digitizing a selected portion of a video image which has been recorded on the video disk. Digitizing is done in a vertical scan, so that 6-bit words are read by the computer at the video line rates, 15750 Hz. At the conclusion of a field, the digitization is resumed at the next vertical column which is

---

* On leave from Soreq Nuclear Research Center, Yavne, Israel.

$(5\text{MHz})^{-1} = 200$ nS to the right of the previous column. The interface assembles three 6-bit pixels into an 18-bit computer word before the transfer to the I/O bus occurs.

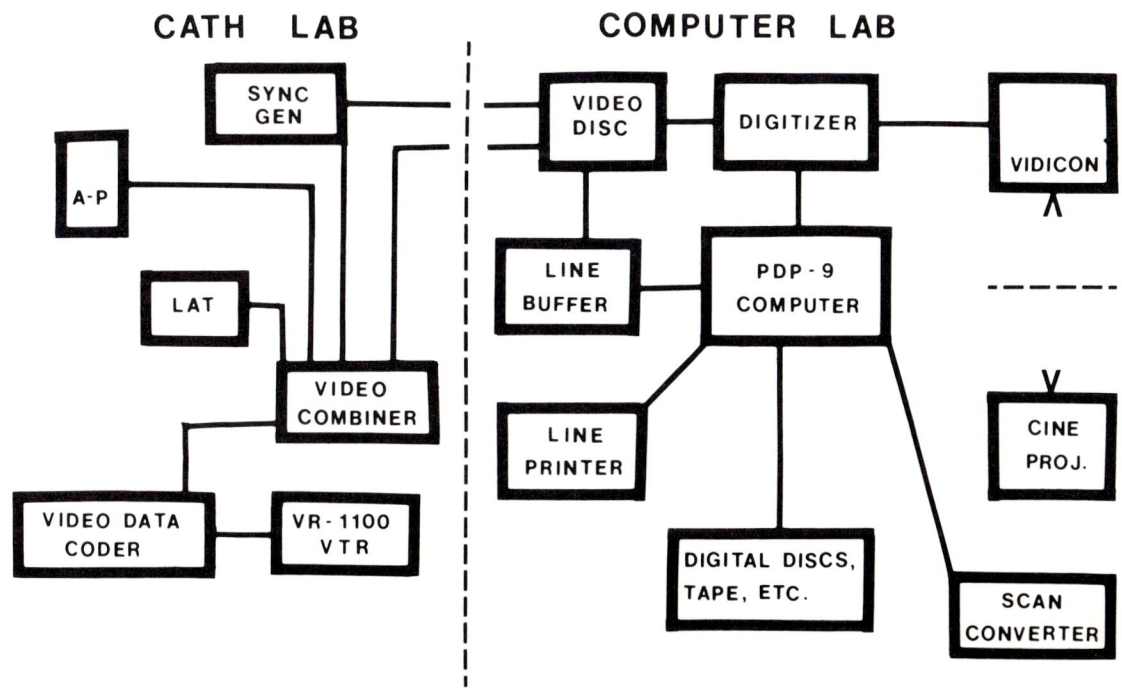

Fig. 1. Block diagram of data analyzing system.

The rectangle defining the region to be digitized is set manually via thumbwheel switches and overlaid on the monitor image. Other functions of the interface include reading the values of the cursors and the track number and stepping the video disk heads.

Conversion of digital data to video format is done in two ways. A video scan converter is interfaced to the computer using 10-bit DACs for the x, y, and z axes in a random address mode. Small amounts of hum and noise in the deflection circuitry and nonlinearities in the DACs have precluded generation of high quality images with a scan converter system. These problems have been overcome by the construction of a video line buffer (VLB) interface.

The VLB contains a 542 x 10 bit bipolar memory with a 100 nS cycle time. This memory can be loaded in a random fashion by the computer and then read out at video rates at a selected video line and field. The burst of data is routed through a high speed DAC to the video disk, where the data are recorded a line at a time. Presently the 5 MHz track on the disk supplies the clock signals, yielding about 250 pixels per line, although the system is designed for clock rates to 10 MHz. About 8.5 sec is required to write an entire video frame of 512 lines. Since fields can be addressed individually, we are exploring the possibilities of stereoscopic viewing by writing left eye data on one field and right eye data on the other, and viewing the monitor through a synchronous shutter.

When the video disk is replayed for digitization and/or calculation, a method is needed to allow an operator to designate a specific point or area of the video picture for special treatment by the computer. We accomplish this by using a specially-designed joystick cursor controller connected to a Tektronix 4013 display terminal. The terminal, operating under control of the computer program, can display ahphanumeric characters and graphics and can send to the computer the coordinates of the cursors positioned by the joystick. A video camera continuously views the terminal screen, and a mixer allows us to overlay the characters, graphics and cursors onto the disk output. A monitor displaying this mixed signal is then used by an operator to trace around a particular object in the picture with the cursors for the computer while the computer draws a line on the screen behind the cursors as they are moved. The computer then recognizes the area outlined in this manner as corresponding to specific elements in the array of the digitized video picture in core, and can do calculations on only those elements as programmed. Any number of such areas can be outlined. One can also control the program by selecting options with the cursors displayed on the screen in menu fashion. A mechanical arm allows the camera to be moved away from the front of the display terminal for use of the keyboard and to swing back easily when needed.

Medical Applications

Our system as described is versatile enough that it can be put to a wide variety of uses. However, it has only recently been completed, and we have had time to apply it to only several types of tasks. More are planned. Some of the present uses are described below.

1. Heart Chamber Volume Calculations.

Video data acquired through the system are analyzed by different methods, thus yielding values of useful parameters, such as chamber volume, ejection fraction, shunts, blood flow or wall motion measurements.[6-13]

When a patient is catheterized for diagnosis and evaluation of heart disease, a small amount of radio-opaque dye is injected into his heart through a catheter inserted into a vein or artery in an arm or leg. As the dye flows along with the blood through the various heart chambers, the x-ray images of these from two orthogonal directions are recorded on movie film and video disk. When the disk is replayed, the heart chambers can be clearly seen on the monitor.

a) *Heart Chamber Volume by Geometric Calculations*. With the disk in single-frame mode and using the joystick-and-display-terminal technique, the cardiologist can trace around each heart chamber. A program accepts the coordinates, interpolates between them and then uses a model to calculate the volume of each chamber. In this model the heart chamber is assumed to be representable as a stack of elipses, the semi-major and semi-minor axes of each of which are the interpolated widths of the tracings in the two orthogonal x-ray views. Simpson's Rule is then employed to integrate the volume of the model, and empirically determined regression equations give the chamber volumes. Results are printed on the screen by the computer. By tracing the images of the ventricles from the video frames corresponding to the end of diastole and the end of systole in the heart cycle, the volume, ejection fraction and other clinically important quantities can be calculated. Since the procedure uses video images rather than the movie frames which must be chemically processed, it can be used to get results quickly while the patient is still on the catheterization table. Also since it does not require digitizing a video frame, the only interaction with the computer is through the two 1200 baud lines of the display terminal. Therefore, any of several computers can be used.

b) *Heart Chamber Volumes by Videodensitometric Calculations*. In these studies we concentrated mainly on left ventricle volume calculations.

The analog data are obtained from the video camera during and after injection of the opaque material, as described above. The material is injected either in the left atrium or the pulmonary artery, thus enabling homogeneous mixing in the blood before entering the left ventricle. Then, a uniform concentration, c, throughout the chamber is assumed in the subsequent calculations.

Before injecting the opaque material, the "background" intensity of the recorded radiation is:

$$I_b = I_0 \exp(-\mu_b x_b) \tag{1}$$

where I_0 is the initial radiation, μ_b is the absorption coefficient of the tissue, and x_b is the thickness of the tissue. If we neglect the variation of the absorption coefficients along the path, due to the change in energy spectrum, the intensity after the injection will be:

$$I = I_b \exp(-\mu c x) \tag{2}$$

where μ is the absorption coefficient of the opaque material at the appropriate energy, c is the concentration, and x the width of the chamber in the direction of the radiation. A measurement of the chamber before the injection timed with the cardiac cycle, provides us the values of I_b at each point. Moreover, since the zero setting of the electronic system is arbitrary, a measurement of zero reference (from thick absorber in the field) is mandatory, and may be obtained by inserting a lead standard on each frame.

Thus, the experimental values at each point yield:

$$\log I_i - \log I_{bi} = -\mu c x_i = D_i \tag{3}$$

$$\sum -D_i = \sum (\log I_i - \log I_{bi}) = -\mu c \sum x_i = -A\mu c V \tag{4}$$

where A is a geometrical (area) coefficient and V is the volume of the chamber. From this value, the ejection fraction or shunt parameters may be obtained through further calculations, without calibrations (see Section 2). However, when the absolute value of

V is needed, a calibration can be achieved by several approaches:

(i) assuming circular cut shape of the aorta and the same concentration of the opaque material over the ventricle and aorta.

(ii) a perpendicular image of the chamber obtained in a biplane system provides the calibration of the maximum values of D_i.

(iii) assuming circular cut shape of the ventricle one can calculate the concentration of the opaque material.

2. Ejection Fraction Calculations.

Calculations of the ejection fraction by analysis of one cardiac cycle can be obtained by:

$$p = EF = \frac{EDV - ESV}{EDV} = 1 - \frac{ESV}{EDV} = 1 - R \tag{5}$$

where EDV and ESV are the ventricular volumes at end diastole and end systole respectively. Using the values for $\sum D_i$ obtained from our data, the ejection fraction will be:

$$p = 1 - \frac{\sum D_i (ES)}{\sum D_i (ED)} \tag{6}$$

In cases in which back-flow occurs, i.e. in aortic valve insufficiency, in order to evaluate the real ejection fraction we need the values of $\sum D_i$ in at least two consecutive cardiac cycles. An indication of back-flow is when:

$$\sum D_i (ES_1) < \sum D_i (ED_2) \tag{7}$$

Since the ejection in two consecutive cycles is equal, we have:

$$\sum D_i (ED_1) - \sum D_i (ES_1) = \frac{c_1}{c_2} \left\{ \sum D_i (ED_2) - \sum D_i (ES_2) \right\} \tag{8}$$

where c_1, c_2 are the concentrations in the two cardiac cycles. In the cases where backflow exists, the real ejection fraction, p', which describes the net amount of blood ejected into the aorta (net forward flow), is calculated by:

$$1 - p' = \frac{c_2}{c_1} = \frac{\sum D_i (ED_2) - \sum D_i (ES_2)}{\sum D_i (ED_1) - \sum D_i (ES_1)} \tag{9}$$

Figure 2 (a and b) shows two digitized frames at the end-systole and end-diastole in density form after background point-by-point subtraction. The difference of the total density calculated at those two points of the cardiac cycle is proportional to the stroke volume, and the total density at the end-diastole is proportional to the left ventricular volume. The calibration can be done by one of the different ways explained above.

Figure 3a is a printout of the density difference between end systole and end diastole. The total density sum is proportional to the stroke volume. Figure 3b is the same printout for the next cardiac cycle. The ratio of the total density sums in the two cycles provides the net ejection fraction, according to equation 9.

3. Lung Studies

Video densitometric techniques provide an unique tool for studying the hemodynamics of the central circulatory system. Bolus injections of radioopaque materials may be sampled at the rate of 30 samples per second from either extended regions of the heart and lung fields or from regions as small as individual vessels. Indicator dilution curves derived from regions of interest chosen appropriately over heart chambers or selected major vessels such as the pulmonary artery or aorta can yield definitive information on the presence and magnitude of intra cardiac shunts.

Figure 4 (a and b) shows 1 of 600 image pairs (anterior and lateral) collected over a 20 second interval after a pulmonary artery injection of contrast material. The images showing the opacified pulmonary artery could be identified. The pulmonary artery is the optimum location for detecting recycled tracer material resulting from left-to-right shunts. The poor spatial resolution of current radioisotopic imaging devices precludes precise delineation of the pulmonary artery.

It has been shown from point sampling with radioactive sources that the pulmonary perfusion pulse can be measured from the changes in the amount of transmitted radiation as it varies with lung perfusion volume during the cardiac cycle. Lung scans with radioactive particulates provide comparable information, albeit with lower spatial resolution than may be achieved with gated videodensitometry techniques. Quantitative videometric and videodensitometric techniques in principle can be used to generate a complete image of the magnitude of the regional pulmonary pulse for the entire lung field. This "functional" image would be derived from averaged images taken at end-diastole or any other phase of cardiac cycle.

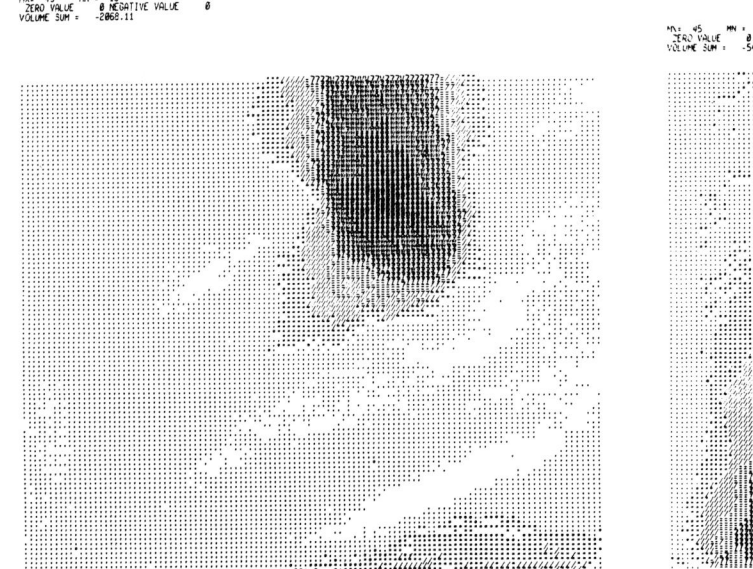
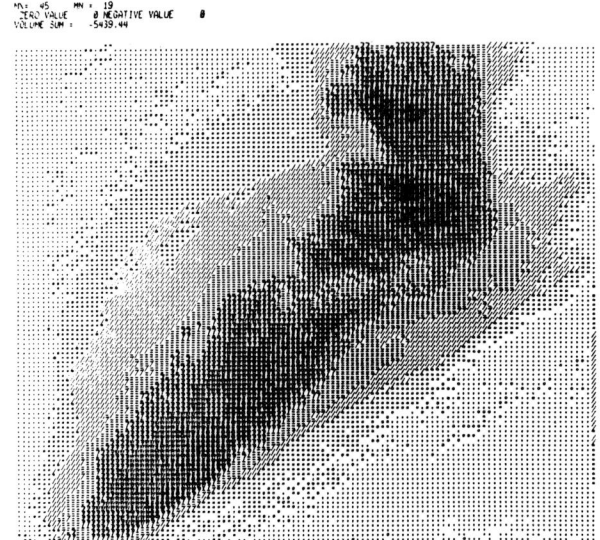

Figure 2. End-diastole (a) and end-systole (b) density images in one cardiac cycle.

4. Analysis of Small Cross-Section Vessels.

Vascular defect analysis has been a meaningful field of interest for blood flow insufficiency diagnosis. In many cases, the small size of the vessels was an obstacle in the analysis of the defects. However, using a video system and special optics, a sufficient magnification can be obtained from the cine-camera recordings. Additional magnification can be achieved during the study by geometrical enlargement, when only the significant area is recorded. Enlargement of x50 has been done to a point where the film grain started to appear. No distortion has been found. Calibration was done using a grid of wire 1 cm apart.

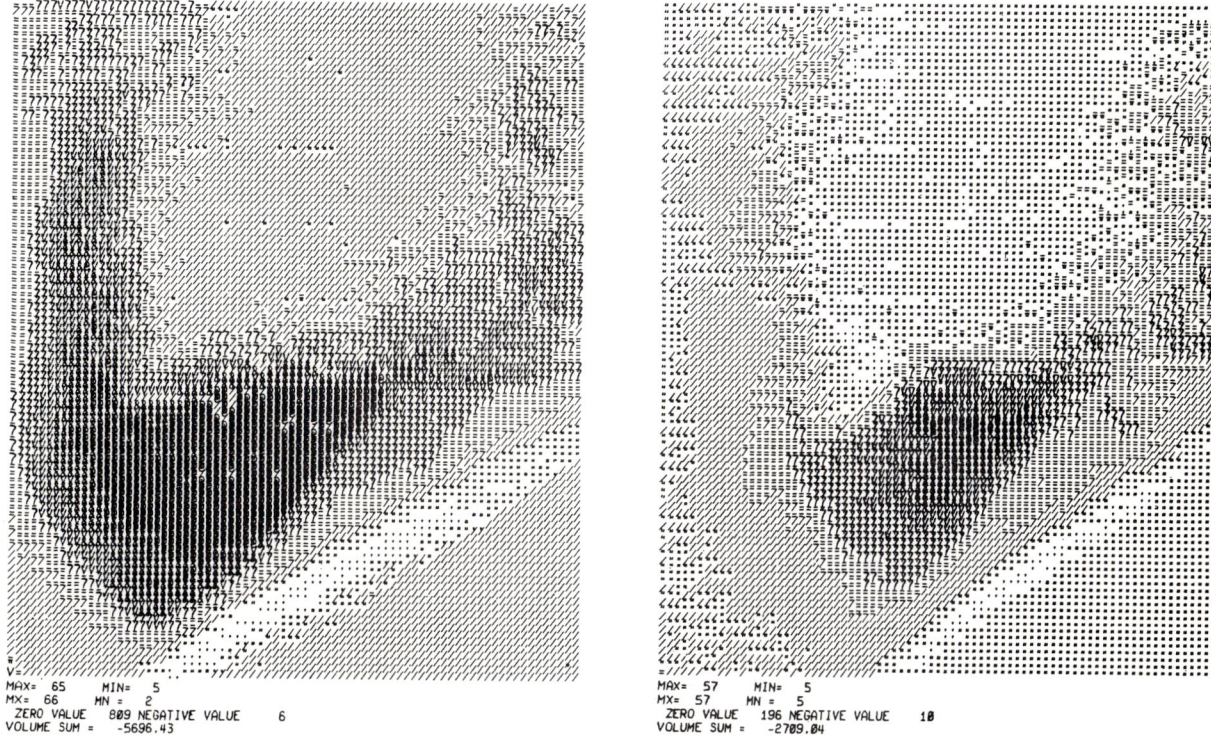

Figure 3. Density differences between end-diastole and end-systole in two consecutive cardiac cycles.

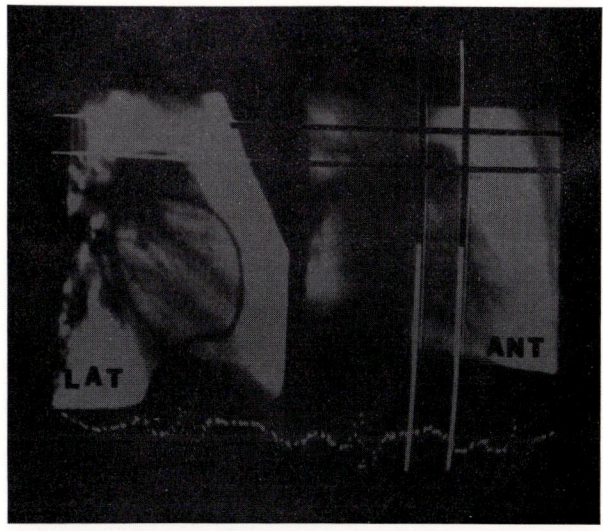

Figure 4a. Redisplay from the video disk of merged anterior and lateral lung frames after an injection of contrast material in the pulmonary artery. Up to 600 merged video frames are stored on the video disk corresponding to 20 seconds of real time data acquisition. Video densitometric curves are generated from region of interest cursors placed over selected regions.

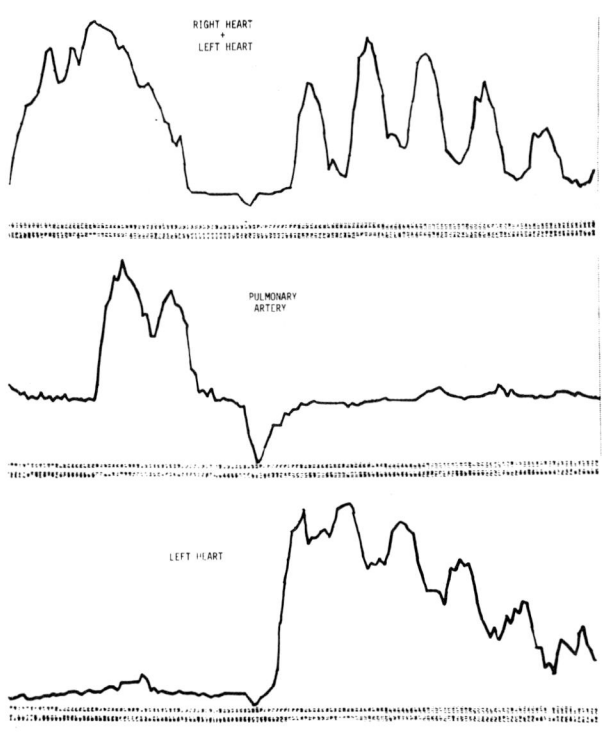

Figure 4b. Video densitometric curves from right atrial injection of contrast material (top) region of interest over total heart, (middle) region of interest over pulmonary artery and (bottom) region of interest chosen only over the left atrium and ventricle.

Conclusions

Measurement of the ejection fraction by the videodensitometric technique is expected to have a higher accuracy than provided by the geometric techniques. This is because the shape of the left ventricle at end-systole is far from cylindrical symmetry, which affects the accuracy of ejection fraction calculations.

For better accuracy, chamber volume calculations can be performed by both methods, namely the geometric and videodensitometric methods (1a, 1b). The geometric calculations can be performed from recordings in which the concentration of the opaque material is higher, which results in a high contrast and better tracing of the chamber's edges. On the other hand, the videodensitometric calculations are performed on recordings obtained after one or several cardiac cycles, when the concentration of the opaque material is reduced. In that situation, the absorption of radiation at each analyzed point is uniquely related to the amount of opacified blood, with no saturation effects. This condition is met by proper choice of the x-ray exposure factors. Thus, the two volume measurements calculated from the same catheterization recordings enable us to obtain a more accurate value of the chamber volume.

References

1. Wood, E.H., Sturm, R.E., Sanders, J.J., "Data Processing in Cardiovascular Physiology with Particular Reference to Roentgen Videodensitometry," Proc Mayo Clin 39:849, 1964.

2. Heintzen, P., "A Simple Method for Recordings of Radioopaque Dilution Curves During Angiocardiography," Amer Heart J 69:720, 1965.

3. Heintzen, P., Bursch, J., "Methods for the Recordings of Radio-opaque Dilution Curves during Angiocardiography," Proceedings of the Association of European Cardiologists 2:39, 1966.

4. Rutishauser, W., et al. "Evaluation of Roentgen Cinedensitometry for Flow Measurement in Models and in Intact Circulation," Circulation 36:951 1967.

5. Bursch, J., Johns, R., Heintzen, P., ed., "Validity of Lambert Beer's Law in Roentgen Densitometry of Contrast Material Using Continuous Radiation," Roentgen-, Cine-, and Videodensitometry, 81, 1971.

6. Chapman, C.B., et al, "Use of Biplane Cinefluorography for Measurement of Ventricular Volume," Circulation 18:1105, 1958.

7. Dodge, H.T., et al, "The Use of Biplane Angiocardiography for the Measurement of Left Ventricular Volume in Man," Am Heart J 60:762, 1960.

8. Ritman, R.L., Sturm, R.E., and Wood, E.H., "Comparison of Volume of Canine Left Ventricular Casts and Angiograms Using Biplane and Monoplane Roentgen Videometry," Physiologist 13:294, 1970.

9. Trenholm, B.G., et al, "Computer Determination of Left Ventricular Volume Using Videodensitometry," Med Biol Eng 10:163, 1973.

10. Ritman, E.L., Sturm, R.E., and Wood, E.H., "Biplane Roentgen Videometric System for Dynamic Studies of the Shape and Size of Circulatory Structures, Particularly the Left Ventricle,: Am J Cardiol 32:180, 1973.

11. Trenholm, B.G., et al., "Automated Ventricular Volume Calculations from Single Plane Images," Diagnostic Radiology 112:299, 1974.

12. Rhea, T.C., et.al., "Ventricular Volume Determination "On Line" Using a Video/Computer System," Circulation 50, 1974, Supplement III (abstract).

Acknowledgments

SCOR in Hypertension, NHLI-HL14192, HL14454-041A1, American Heart Association, and NASA Contract NAS8-30894. IAEA Fellowship SC/202/ISR/7505. Assistance from Hamamatsu Corp. is gratefully acknowledged.

RESULTS OF RATIO TEMPERATURE THERMOGRAPHY

E. L. Dereniak, H. Roehrig, and W. L. Wolfe
Optical Sciences Center, University of Arizona
Tucson, Arizona 85721

Abstract

Our present understanding of breast cancer indicates that increased metabolic activity occurs and thus produces a local temperature increase. This paper evaluates present thermographic techniques used to detect these temperature increases and examines the problem of skin emissivity variations that can produce erroneous temperature measurements. These false temperature variations are on the order of the decision level used by radiologists, and therefore they can cause significant confusion in the interpretation of the thermogram. The ratio temperature thermograph is shown to reduce the effects of emissivity by measuring the spectral radiance at two prescribed wavelengths and ratioing the results. A dual-channel ratio thermograph was built using state-of-the-art detectors and electronics to prove its feasibility. The ratio temperature thermograph was quantitatively evaluated for small emissivity variations. This evaluation demonstrated the instrument's capability of minimizing emissivity effects. It was also evaluated for the detection of temperature changes.

Introduction

Despite considerable medical research in the control and cure of breast cancer, the mortality rate has not changed in the past half century. Breast cancer statistics are appalling when one considers that there will be 90,000 new cases this year (three out of every 1000 women).[1]

Also, only 53% of the women treated for breast carcinoma by surgery will remain free of disease 5 years after surgery. However, 82% of women treated by surgery when the cancer is still confined to the breast are free of disease 5 years later, as compared to only 25% if the cancer has spread to the axillary lymph node. Therefore, it is well established that patients without axillary lymph node metastases have a greater chance of survival than patients whose cancer has spread beyond the breast.[2,3] In addition, the number of patients with axillary lymph node involvement increases directly with the size of the primary lesion, and lesions too small to be palpable have a lower percentage of axillary lymph node metastases.[4] These data indicate that earlier detection of breast cancer is mandatory for a lower mortality rate in the future.

Research has shown that a malignancy causes increased metabolic activity, which in turn produces heat or temperature changes. If these temperature changes can be detected at an early state of cancer, an early warning system could be developed. Thermograms have been used for several years to detect these temperature changes with reasonable success.

Medical Thermography

Figure 1 shows two thermograms, one of a woman with cancer, the other without. The disease is readily visible through the asymmetry in the thermal pattern of the left picture, while the right picture presents an apparently nonsymptomatic case. Very often, however, symptomatic cases go undetected. This is a very disturbing fact. There are strong indications that presently used thermographic instrumentation is not properly designed to eliminate all possible errors in determining the true temperature profile of the human skin. A better instrument or method for finding small lesions early must be developed that is economical, safe from ionizing radiation, and capable of screening the entire population. This paper concerns the development of a new method in thermography, that of ratio temperature thermography.

The technique applied most commonly today in medical thermography is based on the thermal radiation emitted by the human skin. According to the Stefan-Boltzmann law and the Planck radiation law, the radiance of the skin is a function of the emissivity ε of the skin and its temperature T.

As the detector of the medical thermograph scans across the human body, it is exposed sequentially to different radiating elements of the body, each one considered to be of uniform temperature and emissivity. First, the system views a radiating element of temperature T_1 and emissivity ε_1. Then it views a very short time later another element with temperature T_2 and emissivity ε_2. The differences are ΔT and ΔE, which we will assume to be small enough so that they can be written dT and $d\varepsilon$.

The difference in the radiances of these elements is then

$$M_q = \varepsilon \sigma_q T^3$$

$$dM_q = 3\varepsilon \sigma_q T^2 dT + \sigma_q T^3 d\varepsilon$$

$$\frac{dM_q}{M_q} = 3\frac{dT}{T} + \frac{d\varepsilon}{\varepsilon}.$$

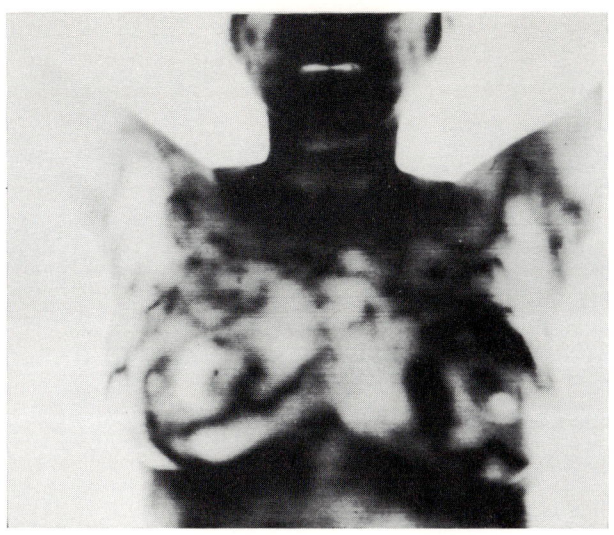 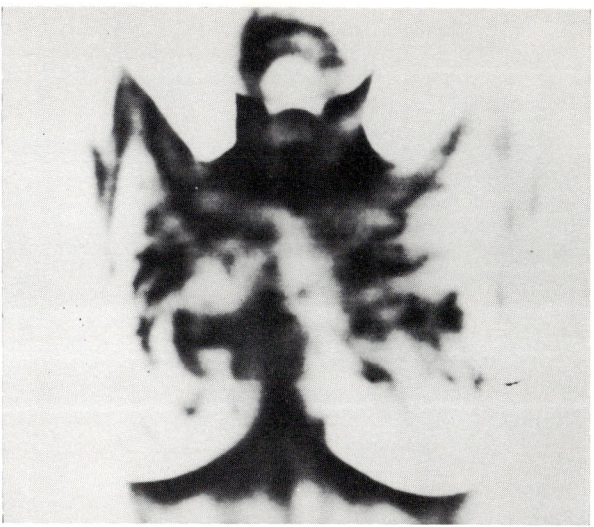

Fig. 1. Thermogram of woman with cancer (left) and woman without cancer (right).

This describes the case for a wideband detector receiving all the radiation in the spectrum, so the Stefan-Boltzmann law applies. Note that the relative irradiance change is due to a change in temperature and emissivity. For the case of a narrowband detector, where Planck's law applies, the relative change in irradiance is

$$M_q(\lambda, T) = \varepsilon 2c\pi(\lambda^{-4})(e^{hc/k\lambda T} - 1)^{-1}$$

$$dM_q(\lambda, T) = \varepsilon 2c\pi\lambda^{-4}(e^{hc/k\lambda T} - 1)^{-2} \cdot e^{hc/k\lambda T} \cdot \frac{hc}{k\lambda T^2} dT + 2c\pi\lambda^{-4}(e^{hc/k\lambda T} - 1)^{-1} d\varepsilon$$

$$\frac{dM_q(\lambda, T)}{M_q(\lambda, T)} = \frac{hc/k\lambda T \cdot e^{hc/k\lambda T}}{e^{hc/k\lambda T} - 1} \frac{dT}{T} + \frac{d\varepsilon}{\varepsilon}.$$

For maximum contrast

$$\frac{dM_q(\lambda, T)}{M_q(\lambda, T)} = 5 \frac{dT}{T} + \frac{d\varepsilon}{\varepsilon}.$$

Figure 2 plots the relative change in irradiance for the two cases. Present interpretation of the irradiance changes is that the emissivity change from one element to the next is zero ($d\varepsilon = 0$). Then, the

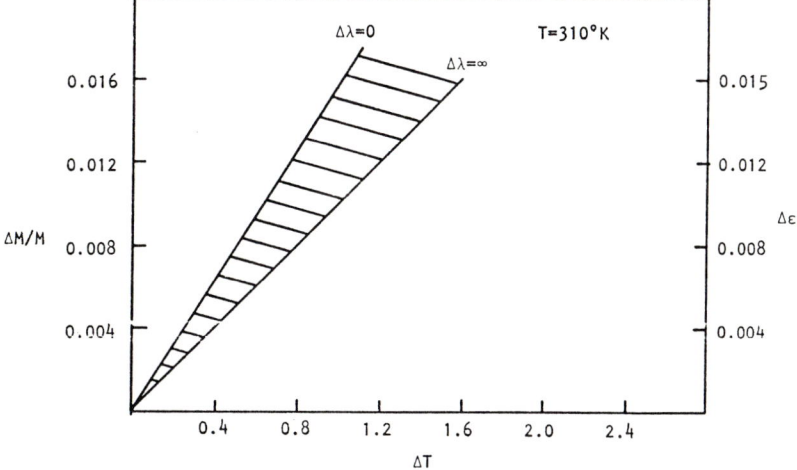

Fig. 2. Monochromatic and total radiation contrast change versus relative emissivity and temperature change.

signal is directly proportional to the temperature difference (dT). However, the emissivity is not constant or is known to only 1% or so. From the equation one can see that a 1% emissivity change causes a 0.75°K difference in the monochromatic case and 1°K difference in the wideband case. A 1°K temperature difference, however, is already considered critical by the physician. There is sufficient literature to indicate that the emissivity change is on the order of 1%.(5)

Ratio Temperature Thermography

Our solution to the emissivity problem is ratio temperature thermography. The concept is shown in Fig. 3. Actually this principle is not new and it is widely applied in illuminating engineering. This principle relates the spectral distribution of the radiation to a temperature. It is well known that the spectral distribution of the radiation of a blackbody is uniquely determined by temperature of the blackbody. This is Planck's law. This applies for a blackbody and also for a graybody (one that has an emissivity smaller than one, which is, however, a constant with respect to wavelength). The spectral distribution can be determined by measuring the radiance at two different wavelength. This figure indicates the emission spectrum of a 310°K blackbody with the emissivity $\varepsilon = 0.98$.

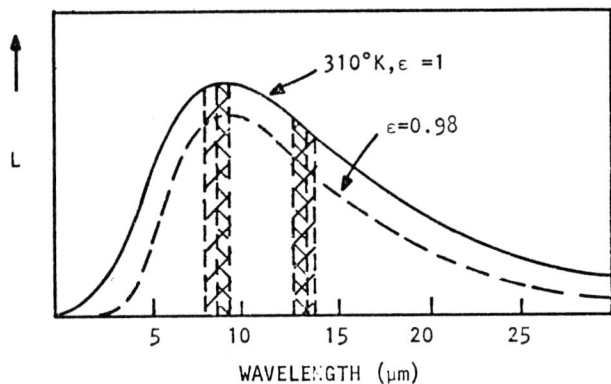

Fig. 3. Ratio temperature measurements for different emissivities (graybodies).

In normal thermography the detector measures the area under the curves, let us say in the 8 to 14-μm region. As the detector scans from an element with $\varepsilon = 1$ to one with $\varepsilon = 0.98$, it displays this difference as the signal.

In ratio temperature thermography, the detector would sample the radiance of the element with $\varepsilon = 1$ at two points in the spectrum, say 9 μm and 13 μm as indicated in Fig. 3 and perform the ratio of these two values. Mathematically, the system output is

$$\frac{D(T,\lambda_1)}{D(T,\lambda_2)} = \frac{\varepsilon_1 C_1 \lambda_1^{-4}(e^{C_2/\lambda_1 T} - 1)^{-1}}{\varepsilon_2 C_1 \lambda_2^{-4}(e^{C_2/\lambda_2 T} - 1)^{-1}} = \frac{C_1 \lambda_1^{-4}(e^{C_2/\lambda_1 T_c} - 1)}{C_1 \lambda_2^{-4}(e^{C_2/\lambda_2 T_c} - 1)}$$

where

T = kinetic temperature of object
T_c = ratio temperature of object
$D(T,\lambda_1)$ = measured irradiance at λ_1
$D(T,\lambda_2)$ = measured irradiance at λ_2.

Solving for ratio temperature

$$\frac{1}{T_c} = \frac{1}{T} - \frac{\lambda_1 \lambda_2}{C_2(\lambda_2 - \lambda_1)} \ln(\varepsilon_1/\varepsilon_2)$$

and the change in ratio temperature is

$$dT_c = \frac{T_c^2}{T^2} dT + \frac{\lambda_1 \lambda_2 T_c^2}{C_2(\lambda_1 - \lambda_2)} \left(\frac{d\varepsilon_1}{\varepsilon_1} - \frac{d\varepsilon_2}{\varepsilon_2} \right).$$

As one can see, the ratio of the two radiance values is proportional to the temperature. For the case when the emissivity is constant with respect to wavelength, the ratio is independent of the emissivity, and the temperature measured is the true temperature of the emitter.

If the emissivity is not constant with respect to wavelength, then the temperature determined from the ratio is different from the true temperature and is called the ratio temperature.

The method of ratio temperature will eliminate the effect of emissivity provided there is a strong correlation between emissivities at wavelengths λ_1 and λ_2. An analysis has shown that the correlation need be only 0.366 in order for the ratio temperature measurement to be just as good as the present brightness temperature measurement.

In order to check this theory, a two-color radiometer was constructed. The instrument used was a modified FLIR system. The detector package in the focal plane was modified to have two channels using a beamsplitter. This arrangement of one optical system and a beamsplitter was chosen to avoid the registration problems that typically occur in arrangements of two optical systems. The beamsplitter assembly is shown in Fig. 4. One channel operated at 5 μm and the other at 13 μm. The actual beamsplitter assembly was made of copper. The beamsplitter itself is sapphire. It transmits energy below 7 μm and reflects energy above 11.5 μm. The liquid helium device that contained the beamsplitter assembly is shown in Fig. 5.

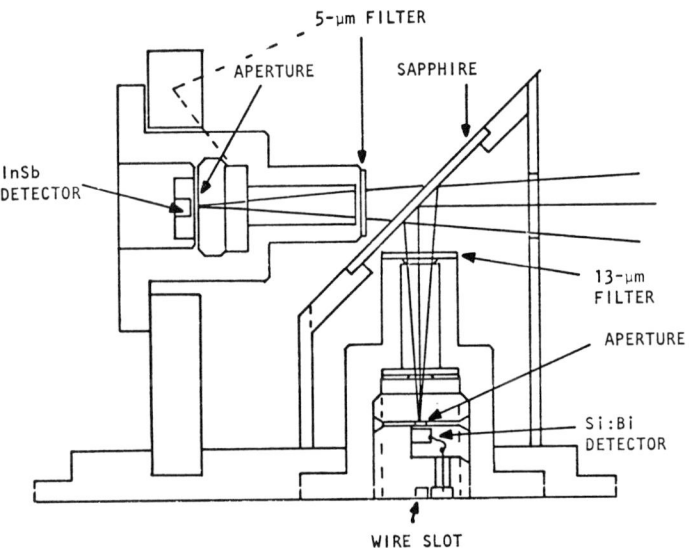

Fig. 4. Optical layout of focal plane assembly.

Fig. 5. Focal plane assembly.

The spectral response of the detectors used is shown in Fig. 6. The InSb is a photovoltaic detector and the bismuth-doped silicon is a photoconductor. The system performance is shown in Table 1. The 5-μm channel and the 13-μm channel had a comparable NETD. The system was checked for sensitivity and spatial resolution using an infrared bar chart similar to the three-bar targets used in the visible region.

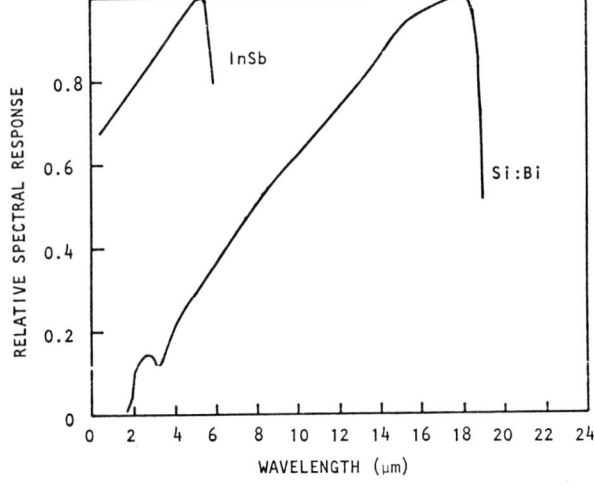

Fig. 6. Detector spectral response. (Detectors donated by Aerojet Electrosystems.)

Table 1. Channel characteristics and performance.

	5-μm Channel InSb	13-μm Channel Si:Bi
D*	8.3×10^{11}	6.6×10^{11}
NEP	8×10^{-14}	1×10^{-13}
$\lambda_1 \rightarrow \lambda_2$	4.5–5.5 μm	12.9–1.4 μm
NETD	0.145°K	0.136°K
Optical resolution	0.75 mm	0.75 mm
Line time	75 ms	75 ms
Frame time (200 lines)	15 sec	15 sec

In order to simulate a change in emissivity, a combination of carbfusion and carbon was painted on a volunteer's arm as shown in Fig. 7. Line scans across the emissivity changes are shown in Fig. 8. The 5-μm and 13-μm channels definitely detect the emissivity change whereas the ratio temperature channel eliminates the emissivity effects. Presently this system is being modified to generate a raster scan so that two-dimensional thermal pictures can be made.

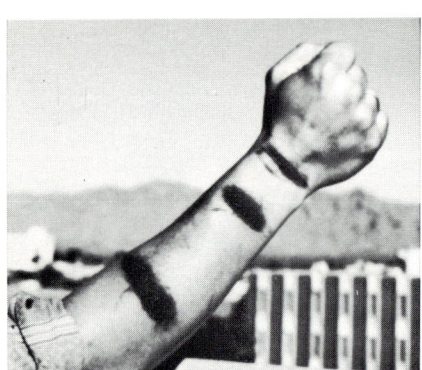

Fig. 7. Emissivity changes on skin.

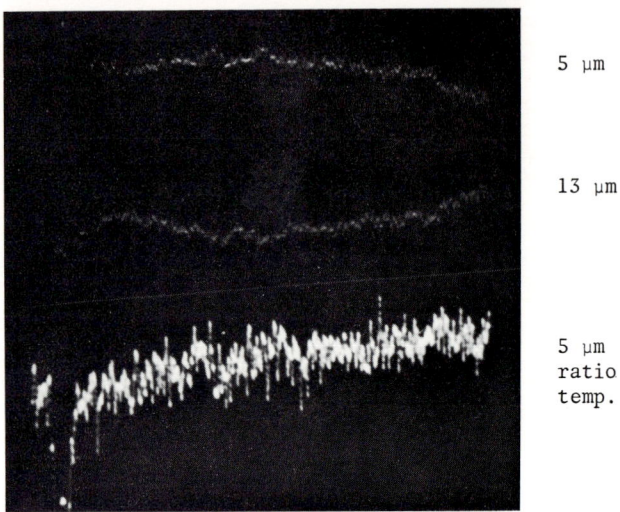

Fig. 8. 5-μm and 13-μm integral temperature and ratio-temperature scans.

References

1. Consumer Reports, "What women don't know about breast cancer," Consumer Reports, March 1974.
2. Fisher, B., Slack, N. H., and Bross, I. D. J., "Cancer of the breast: size and neoplasm and prognosis," Cancer 24:1071, 1969.
3. Freundlich, I., Radiology Department, University of Arizona, private communication, 1976.
4. Freundlich, I. M., Wallace, J. B., and Dodd, G. D., "Thermography and mammography in the detection of early breast carcinoma and as a possible method of population survey," Proceedings of the First International Symposium on the Detection of Cancer, 1968.
5. Steketee, J., "Spectral emissivity of skin and pericardium," Physics in Medicine and Biology 18:686, 1973.

RADIOMETRIC FLIR FOR THERMOGRAPHY

Scott P. Way
Development Engineer
Honeywell Radiation Center
Lexington, MA 02173

Abstract

The Honeywell Medical Scanner has been developed to accurately measure human body temperatures in a laboratory environment. Specifically, it detects abnormal blood profusion, an indicator of cancer, in the breast and surrounding areas of the body. The scanner uses Honeywell's high performance TV compatible FLIR technology, similar to the AN/AAQ-9 and Chaparral FLIRs, to produce radiometrically calibrated FLIR imagery which may be digitized, for computer processing, or stored on video tape for playback at a later date. The calibrated imagery is produced by referencing the image seen by the front objective to two internal standards. These standards set the gain and level of the output video such that voltage out corresponds to a given observed temperature. By sampling a section of the frame and integrating over several frames, thermal resolution of 0.1°C may be achieved.

Introduction

The Honeywell Medical Scanner was designed and built for the University of Oklahoma College of Medicine under a grant from the National Cancer Institute to provide a real time thermal imaging device for the early detection of breast cancer in women. The device permits a technician to perform the scanning function and record the video imagery on tape. In this manner, a doctor can review the tapes at his or her convenience. The video imagery is also radiometrically calibrated to provide absolute temperature readout.

The medical scanner program is divided into 3 separate phases. Phase I, the current phase, is designed to demonstrate the advantages of real time versus stored imagery in speed, sensitivity, and operator convenience. The radiometric usefulness of the device will also be proven. Phase II will use digitized video imagery in a computer to determine if the screening function can be done automatically. In this way, a large percentage of the patients undergoing thermographic testing can be screened out quickly and economically. During this phase of the program, localization techniques and digital processing algorithms will be utilized to standardize and screen prospective patients. Phase III will be to operate the unit in the field, probably out of a van, and prove the capability of the unit with a large sample of the population.

In addition to the components found in a standard Thermal Imager, the Medical Scanner contains several additional components to provide it with radiometric capability. Among these is a special video processor which provides the calibrated video to be stored on tape, and a set of playback electronics to process the video when played back from the tape recorder. Also, the playback electronics module allows the operator to adjust the gain and level of the stored video, and controls are provided to select any portion of the scene whose temperature is desired. These features are discussed in detail in the following sections.

Video Processor Theory of Operation

The medical scanner uses a serial scan technique to produce TV compatible imagery. Since each detector "sees" the full scene and a delay and add technique is used, a single gain and level calibration can be performed at the preamplifier output of each channel.

The Honeywell Medical Scanner uses two precision thermistor targets mounted on, but thermally isolated from, its field stop. In this manner focusing problems are eliminated and thermal instability due to convection currents is minimized.

Figure 1 is a diagram of the field stop showing the relationship between a target in the scene, the field stop, and the two reference targets which provide radiometric calibration of the scanner.

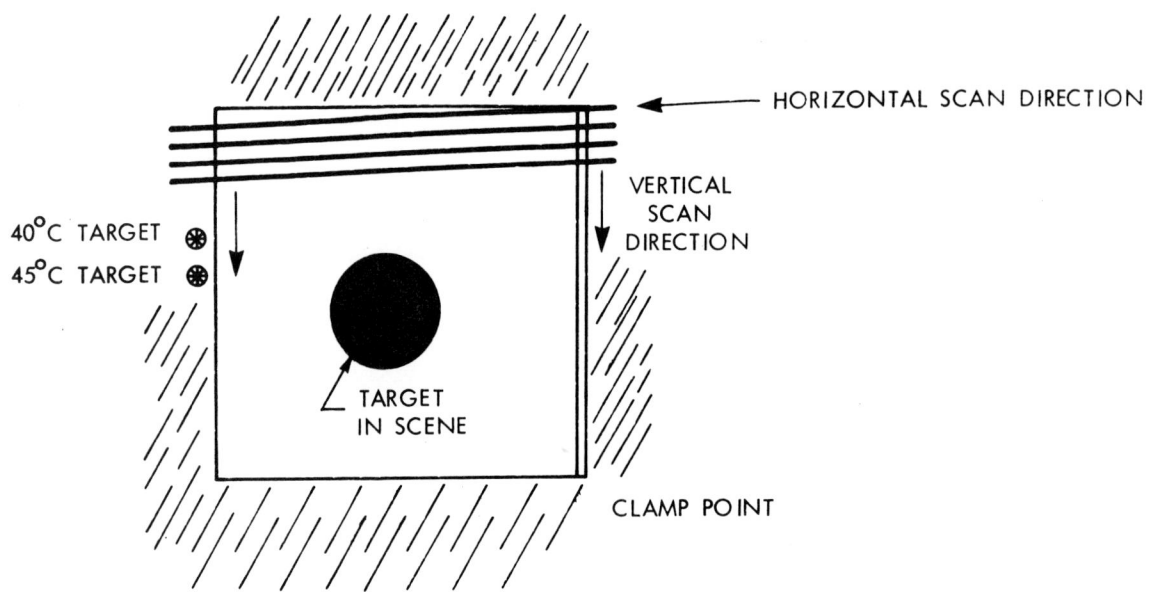

Fig. 1 Field stop showing placement of temperature standards, scan direction and isothermal reference point

The field stop defines the external field of view of the scanner. The vertical and horizontal blanking is adjusted such that the field stop edges occur just outside the unblanked video. This blanking, however, occurs in the last stages of the video processor such that the two reference targets, which are also scanned, may be processed.

Figure 2 shows a portion of the voltage waveform at the preamplifier output. This waveform is applied to the AGC (Automatic Gain Control) circuit whose gain is controlled by sampling the 45°C target and comparing its voltage level to a known reference of 0.1 volt. Next, the video waveform is clamped during the field stop time to a voltage which forces the 40°C target to 0 volt. This establishes two temperatures to two voltage levels and, assuming a linear transfer function through the processing chain, establishes a voltage to temperature curve such that any voltage can be transformed to temperature by use of this curve. Also sync and blanking signals may be added to the existing waveform without disrupting the calibration. Thus, a composite video waveform is produced which contains radiometric information without reference targets in the scene. This sequence is shown in Figure 3.

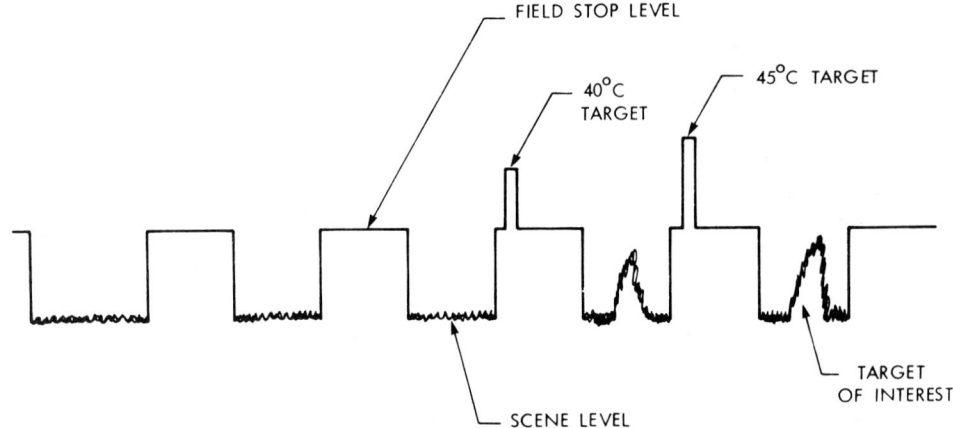

Fig. 2. Medical scanner preamplifier output

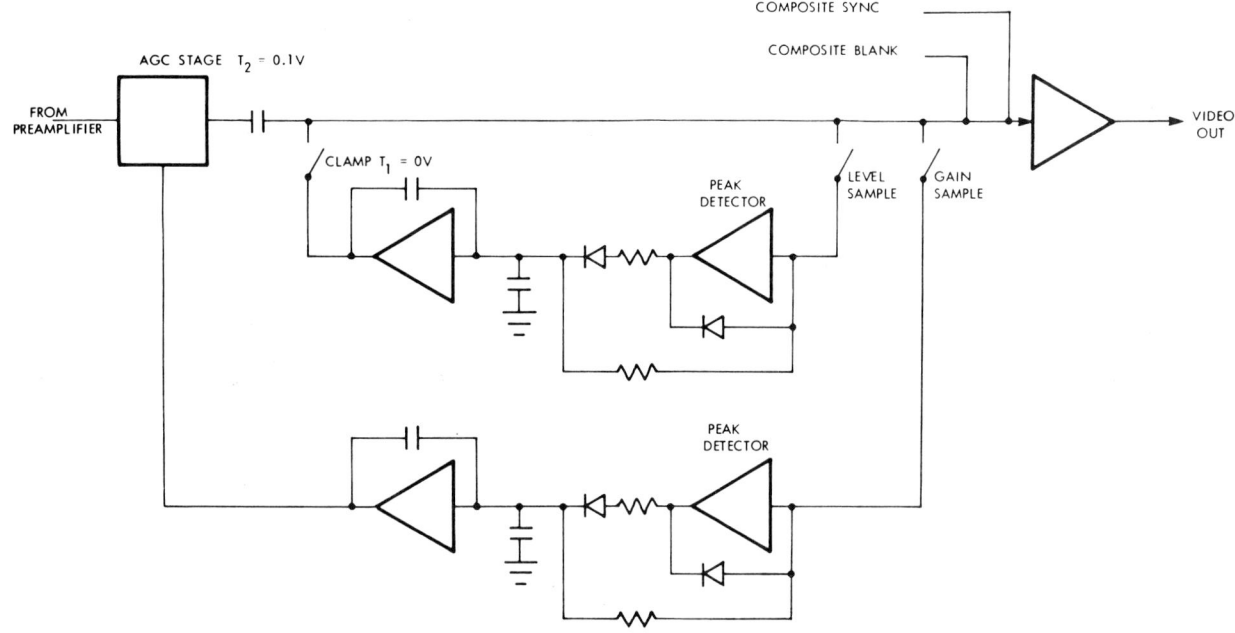

Fig. 3. Block diagram video processor

The video waveform has several advantages. One, it contains wide dynamic range information. Typically, for the medical scanner the video has about 25°C information with approximately 0.2°C sensitivity. This corresponds to 42 dB dynamic range. The temperature range of the video processor is primarily limited by a diode clipping network which ensures that the output of the video processor is EIA standard video. Since the FLIR system is detector noise limited, we can decrease the gain of the video processor prior to the clipping stage to obtain even wider dynamic range information, but less sensitivity. For this application sensitivity is of prime concern. A second advantage is that the output video is television compatible and can be stored on standard video tape. This allows one to utilize a wide variety of commercially available video processing equipment thus decreasing design costs and time for special equipment. The user must be careful, however, that he or she does not distort the radiometric integrity of the video by using nonlinear storage or processing devices. Unfortunately, most analog tape recorders are nonlinear, and correction factors must be applied. This is difficult if one tape recorder is used during the record mode and another is used for playback. However, calibration targets may be recorded as a reference at the beginning of each tape. This technique allows calibration for individual recording machines.

Playback Electronics Theory of Operation

For the medical scanner application, it was felt that temperature differences as low as 0.1°C were necessary to determine the presence of breast assymmetry, the primary indicator of breast cancer in a thermogram. To realize this sensitivity in FLIR imagery, it is necessary to integrate several frames of video. A gated integrator is used which is controlled by the sampling window. The sampling window also controls a second integrator which is driven by a constant current source. This integrator triggers a comparator which samples the first integrator voltage and then resets both integrators. Since this second integrator is charged with a constant current source, its charge-up time to the trigger voltage is constant thus providing a constant integration time to the first integrator regardless of video content or window size. The output of the first integrator is, therefore, directly proportional to the sum of the input video. This sampled voltage is displayed on a D.P.M. which has been calibrated to read directly in degrees centigrade. This procedure is shown in Figure 4.

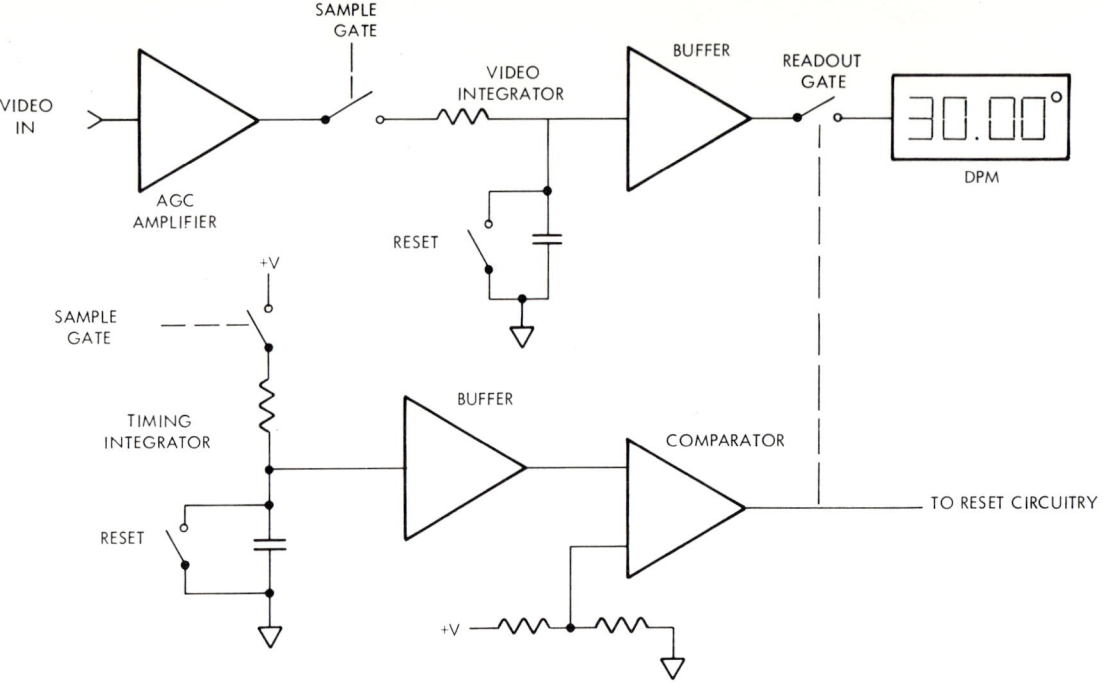

Fig. 4. Playback electronics block diagram

Clinical

The unit shown in Figure 5 along with the playback electronics and tape recorder is currently located at the ACS-NCI Breast Cancer Demonstration Project at University of Oklahoma. Clinical trials by Drs. JoAnn Haberman, Tom Love and James Goin, are being conducted in a specially air-conditioned room which is kept at a constant 21°C. This room allows the patients outer skin layer to cool to room temperature permitting the underlying venous patterns to locally warm the surface. The cool temperature also eliminates false readings due to perspiration. During the initial testing phase of the instrument, it was also noted that the absolute temperature measurements were as much as 0.2°C in error for every 1°C of room temperature drift. Thus the constant room temperature also is necessary to thermally stabilize the unit and insure its accuracy. Normally, the unit is allowed to stabilize for 1/2 to 3/4 of an hour before measuring any absolute temperatures. This insures that all internal components have reached equilibrium, and all thermal drifts have been eliminated. Once this has been done, the scanner is ready for calibration. Calibration is performed by alternately reading the temperatures of two blackbodys and adjusting the gain and level calibrations so that the machine reads within ±0.1°C of either temperature. Care is taken that both targets are in the same spot on the field of view since errors due to nonuniformity at the extreme edges of the field can approach 0.5°C. This effect is believed to be optical in nature and hopefully, can be eliminated in future units. If a mirror is used, as in the case of supine patients, then the unit must be calibrated through the mirrors.

Once the unit has been calibrated, the operating technician may begin screening patients. Normally, patients are brought in and placed in a supine position and a front surfaced mirror is used to project the patients thermographic image to the scanner. Once the patient has been positioned, the operator proceeds to record a few seconds of tape in each of three positions, front, right side and left side. Since the tape may be played back as many times as necessary, twenty to thirty seconds is usually sufficient. Full dynamic range information is recorded on the tape eliminating the need for a technican to set up gain and contrast levels before recording. During playback, the gain level may be set by the interpreter to most prominantly display any features of interest. Typically, all temperature levels between 18°C and 43°C are stored on tape. In computer studies, the full dynamic range will be utilized to detect abnormalities.

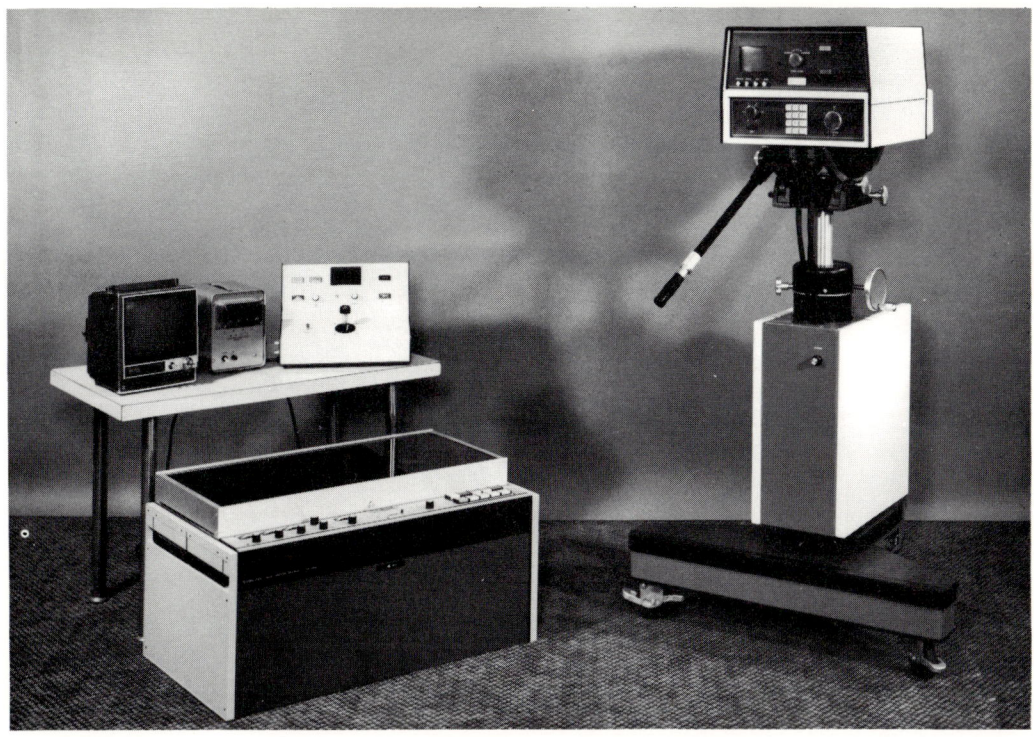

Fig. 5. Honeywell Medical Scanner

Summary

A unit has been built which has demonstrated that it is possible to obtain TV compatible radiometric information. This unit has several advantages over other presently available units in speed, sensitivity, and operator convenience. Further development is necessary to make the unit suitable for the mass screening application in a poorly controlled environment, but it is felt that the technology exists to make this program successful.

Acknowledgement is given for support by the National Cancer Institute under Contract No. 1-CB-43869 and the University of Oklahoma under Subcontract OU-HSC-74-52.

Appendix

Shown below are typical thermograms taken with the Honeywell scanner showing a normal and abnormal patient.

Figure 1 is a thermograph of a patient showing an abnormal pattern. Note the large light area on the right breast, between lines 2 and 3 down from the top, which is the cancerous area. Also, note the very pronounced venous structure on the right breast. This increased blood profusion indicates a growth. In tests, the temperature of that area on the right breast was 1°C higher than the corresponding temperature on the left breast.

Figures 2 and 3 are thermographs of normal patients. The symmetry from breast to breast can be seen here and the absence of any prominent veins indicates a normal pattern.

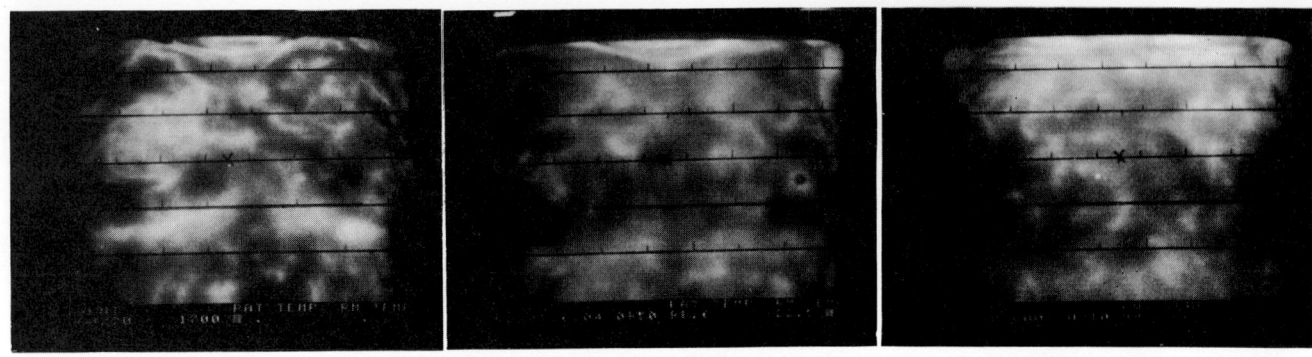

Fig. 1. Thermograph showing abnormal conditions

Fig. 2. Thermograph showing normal conditions

Fig. 3. Thermograph showing normal conditions

A NIGHT VISION AID AS A CONSUMER PRODUCT

James H. Burbo, P. E.
ITT Electro Optical Products Division
Roanoke, Virginia 24019

Abstract

Second Generation Night Vision Devices have received some consideration as commercial products, but rarely as consumer goods, not from lack of potential applications, but because their high cost was inevitably considered to reduce the size of the potential market nearly to the vanishing point. Working with the Retinitis Pigmatosa Foundation, ITT has designed a Second Generation Intensifier Monocular as a medical prosthetic aid for persons suffering from retinal degenerations causing night blindness. The proposed use environment for such an instrument posed an altogether different set of contraints on the design from those encountered in designing military instruments; some requirements of course are relaxed, but others, surprisingly are at least as stringent as those for military application, though often for different reasons. This paper describes the resulting instrument, showing the tradeoffs among cost, performance, human engineering, and environmental considerations, and indicating the rationale behind the design decisions made.

Introduction

The concept of a hand-held, monocular, image intensifier, small enough to be comfortably carried in the pocket, arose fairly naturally with the availability of 18 MM, 2nd generation image intensifiers. The primary impetus for the development of these image tubes, both proximity-focus and electrostatic-inverting types, came, in turn, from the army's requirement for the head mounted binocular systems commonly referred to as night vision goggles. These goggles are just now entering army inventory in significant quantities.

A number of companies have offered monocular or "pocketscope" designs commercially, and several offer quite comprehensive systems, with an extensive selection of interchangeable objectives, monocular and biocular eyepieces, and camera adapters. The market for these has been relatively small, presumably because of the relatively high unit price.

Background

The first pocketscope sold commercially* by the ITT Electron Tube Division (now Electro-Optical Products Division) went to the Western Blind Rehabilitation Center, at the Veterans Administration Hospital in Palo Alto, California where it was used in mobility studies with patients suffering from night blindness.[1]

*Model 4909-18 S/N 9002, February 1973

Figure 1. Type 4909-18 Pocketscope, Predecessor to the Night Vision Aid

Figure 2. The First "hard" mock-up of the Night Vision Aid

Soon thereafter we were placed in contact with Dr. Eliot Berson of the Department of Ophthalmology, Harvard Medical School, and Massachusetts Eye and Ear Infirmary, Boston. Dr. Berson had been working on retinitis pigmentosa, a hereditary degenerative disease of the retina of the eye whose earliest symptom is night blindness, and had done experiments which indicated the possibility that an image intensifier device could be designed which would be of considerable utility to patients suffering the early stages of the disease.

Retinitis Pigmentosa

Retinitis pigmentosa is a genetically determined degenerative disease affecting the retina of the eye. Its first symptom is night blindness, followed by progressive loss of peripheral vision. Its most usual onset comes between six and twelve years of age, but both the time of onset and the rate of progression of the disease vary widely among patients. Its usual inheritance pattern is autosomal recessive, with approximately one person in eighty carrying the recessive gene. It occurs at a much lower incidence as an autosomal dominant and a sex-linked recessive. Given the gene pool mentioned above, there are estimated to be between 100,000 and 200,000 persons in the United States suffering from the disease. Whether the incidence is the same across racial and geographical boundaries is uncertain.

Research on this disease has been at a rather low level until quite recently. With the founding of the National Retinitis Pigmentosa Foundation in 1971, and their sponsorship of the Berman-Gund Laboratory for the study of retinal degenerations at the Massachusetts Eye and Ear Infirmary in Boston (opened April 1974), this situation has improved considerably.

Clinical Testing

Proof of feasibility. The experimental work that showed that persons with night blindness could profit from the use of an image intensifier instrument was done by Dr. Berson with a three power first generation scope[2] (GTE Sylvania 221) and a unity power second generation scope furnished him by the U. S. Army Night Vision Laboratory.[3]

He was later supplied with a small number of model F4909 pocketscopes with varying gains and fields of view and made further studies aimed at more closely defining the parameters of the optimum instrument for the purpose.[4] Over forty patients with symptoms ranging from mild to severe used two or more of the scopes each for periods of three to five days, and reported the usefullness and limitations of the instruments used. These reports, while necessarily subjective, were combined with clinical findings on the severity of the patient's visual disability and laboratory measurements of visual thresholds of these patients with and without the night vision device. The resulting user profiles developed the following information which was used in setting up a device design trade-off matrix:

1. The use environment is predominantly urban. The field of view will contain various light sources most of the time.

2. Average period of use is suprisingly short - only a few minutes. Few patients ever used the device more than 15 minutes at a stretch and none reported more than 2.5 hours of on-time in any one night.

3. Use of a wide angle objective (80° in the tests) with consequent demagnification was of considerable help to a few patients with severely restricted visual fields of view.

4. The required instrument gain, defined as the viewing screen luminance divided by the scene luminance, is relatively modest compared to that required in military situations. Instrument gain as low as 200 is quite satisfactory for many patients. The screen saturation luminance, on the other hand, should be as high as a reasonable expectation of a good tube life will allow. (Tube life is a direct function of light level within certain limits)

5. Cost of the instrument will be of vital importance.

Design Phase

Trade-Off Analysis

Method. Based on the above information, and knowledge of image intensifier technology, a trade-off analysis was made, in which six possible first generation mechanizations, five second generation mechanizations, and three hybrid mechanizations were compared for size, weight, cost, gain, resolution, tube life, maximum screen luminance, degradation of performance by bright lights in the field of view, maturity of technology, and availability of components. (Table 1) In the course of this analysis, "paper" designs of each mechan-

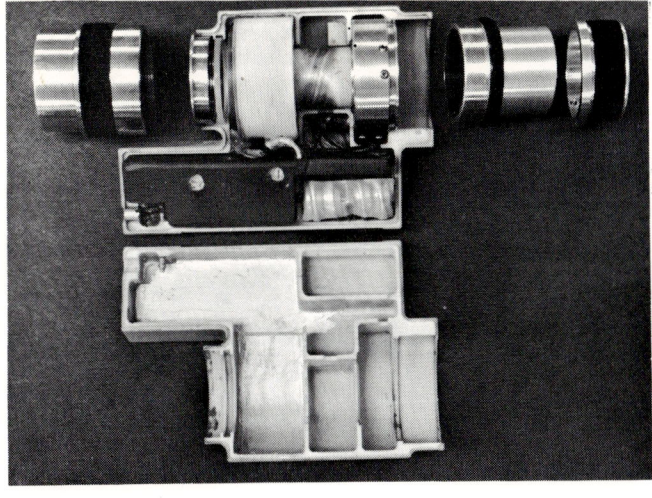
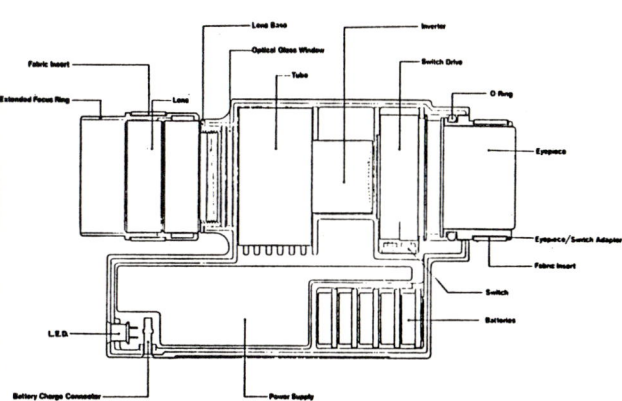

Figure 3. First Prototype just prior to final assembly

Figure 4. Component Layout of the Night Vision Aid

ization were carried as far as necessary to supply the information needed for meaningful comparisons. No hardware was built, though some short additional experiments were run with existing generation I and generation II equipment. We follow the conventional terminology whereby generation one tubes are cascadable modules having gain due only to electron energy increase, and generation two tubes are characterized by having gain due to electron multiplication in a microchannel plate.[5] (Table 2)

Results. The large weighting factor on high light level performance dictated by the urban use environment was almost solely responsible for the elimination of all five of the first generation designs. Two second generation and all three hybrid designs were ruled out by various combinations of adverse factors the more important being cost, size, and weight. This left three designs, all 2nd generation to go into the next design stage, that of three-dimensional mock-up fabrication by human factors consultants.

Envelope Design. At this point the three viable design alternatives were:

1. Gen II Electrostatic Inverter.

2. Gen II Wafer with Integral Fiberoptic Twist.

3. Gen II Wafer with "paste-on" Fiberoptic Twist.

The latter two are close enough in size to each other that we felt justified in asking our industrial design consultants* to do only two mock-ups, one for the Electrostatic Inverter tube (which is the larger) and the other sized to accept either of the wafer designs.

Twenty-six styrofoam models (essentially "three-dimensional sketches") were built, within the constraints that they provide: (1) The correct space for the optical trains composed of the tube configurations outlined above, an eyepiece lens of approximately 25 mm focal length, and an objective lens mount which would accept a standard C-mount lens; and (2) That they provide a certain minimum volume for the batteries, power supply and switch. From among these, one design was selected for further development. Two more detailed (and rugged) models of this configuration were built (photograph - Figure 2), one in each size, and after very minor modifications, the exterior configuration design was frozen. At this point the decision was taken to drop work on the larger configuration based on the electrostatic inverter tube and concentrate on the wafer tube designs.

Specifications of Instrument at Design Freeze Point.

    Configuration: Monocular, self-contained, hand held
    Magnification: 1 x
    Field of view: 40°
    Gain: 250 to 800 factory set per R_x

*Human Factor/Industrial Design Inc., New York. Mr. Douglas Spranger was in charge of this design effort.

Resolution:	0.5 line pairs/milliradian minimum
Automatic brightness control:	yes
Bright source protection:	yes
Focus range:	20 cm to infinity
Diopter range:	factory set per R_x
Power source:	rechargable battery
Operating Temp. range:	$-18^\circ C$ to $40^\circ C$
Moisture:	100% humidity and rainproof
Image tube:	18mm Gen II with F.O. twist
Objective lens:	25mm F.L. F/1.5 C-mount
Eyepiece:	25mm F.L. 40° FOV
Auxiliary illuminator:	gallium arsenide phosphide LED

TABLE 1
Performance Factors & Weighting

FACTOR	WEIGHT	COMMENTS
cost of tube	1.0	--
cost of P.S.	.1	--
size	.2	--
weight	.15	--
gain	--	Go-No-Go decision only
resolution	.05	Patients vision usually limiting factor
tube life	.5	--
max. brightness	.1	--
high light level performance	.5	--
other	.2	Distortion, technological uncertainties, component availability

TABLE 2
Mechanizations considered - All are 18 MM input and output format

TUBE GENERATION	TUBE CONFIGURATION	IMAGE ERECTION
I	3 stage Electrostatic Inverter	tube
I	2 stage Electrostatic Inverter	prism
I	2 stage Electrostatic Inverter	lens
I	2 stage, 1 E.S. Inverter, 1 Wafer	tube
I	2 stage Wafer	prism
I	2 stage Wafer	lens
II	Electrostatic Inverter	tube
II	Wafer	fiberoptic twist (integral)
II	Wafer	Fiberoptic twist (external)
II	Wafer	prism
II	Wafer	lens
II & I	Wafer (II) & E.S. Inverter (I)	tube
II & I	Wafer (both)	prism
II & I	Wafer (both)	lens

Development Cycle

The first hand-built prototype unit was finished in November of 1974, and donated to the National Retinitis Pigmentosa Foundation. Subsequent prototypes underwent engineering field tests and a redesign cycle was initiated in order to achieve a final design that was producible on an assembly-line basis, at minimum cost. In the cycle several significant alterations were made in major components (aside from the image tube itself which has been discussed above).

Eyepiece Lens

The eyepiece lens used on the night vision goggles (AN/PVS-5) has superlative performance but was considered to be too expensive for this application. A number of in-

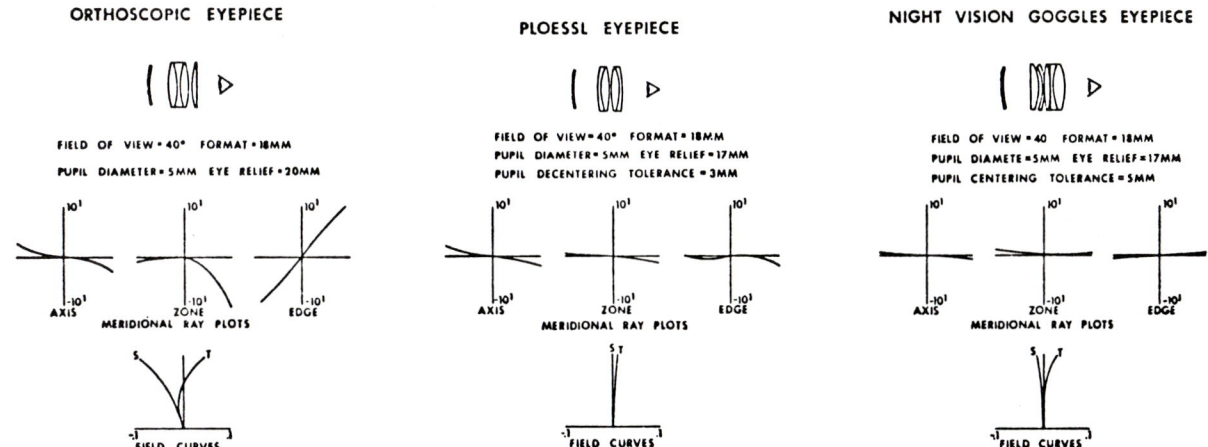

Figure 5. Compromises in Eyepiece lens design.
The ray plots and field curves are from computer ray traces.

expensive eyepieces of the Kelner, Erfle, and Orthoscopic designs were field tested, but were found to have marginal performance, largely because of very tight pupil centering requirements; i.e. they were too unforgiving with respect to the placement of the eye. A new design was indicated, which would be a compromise between these two extremes, and was produced by Dr. Douglas Sinclair of the Institute of Optics, University of Rochester, acting in a consulting capacity. It is a modification of the Plossel type, consisting of two identical doublets, the aberations being balanced for optimum performance in this application.

Objective Lens

In the case of the objective lens, a wide range of designs having acceptable performance were already on the market at quite reasonable prices since the 16-mm motion picture camera and one-inch vidicon camera both have requirements quite similar to those of the night vision aid. The standard objective is a 25-mm focal length, F-1.4 lens of the modified double gauss persuasion, without iris, in a focusing mount whose exterior configuration only is customized for this application. It is provided by a Japanese supplier. Its mounting interface is the industry-standard c-mount, making it possible to use alternative lenses when desired simply by unscrewing the normal lens and screwing in any c-mount lens which will physically clear the power supply protrusion below the lens mount.

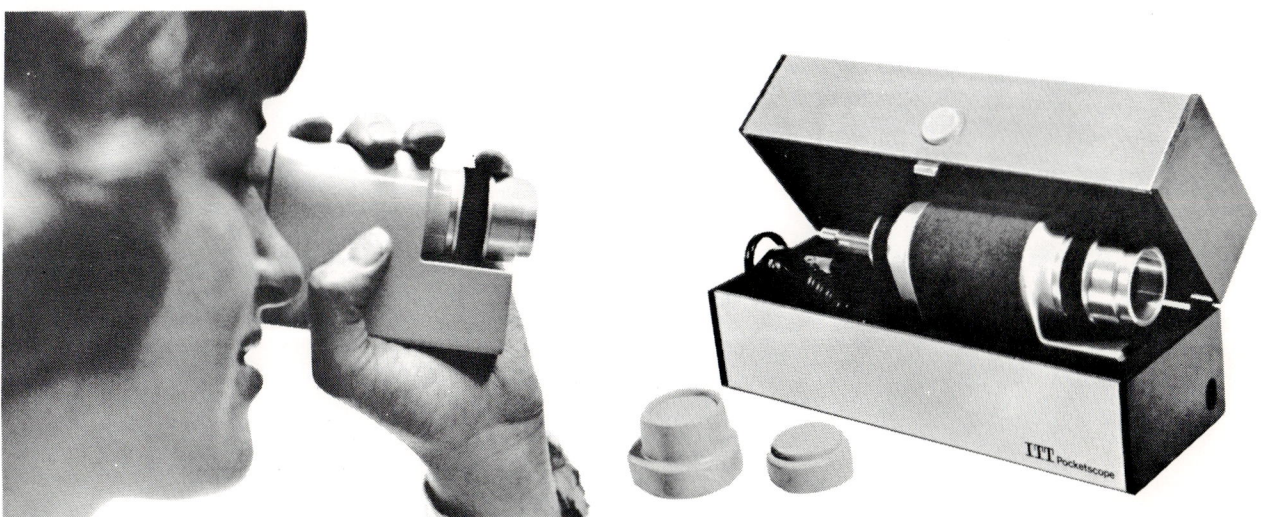

Figure 6. The Night Vision Aid in Use.

Figure 7. The Night Vision Aid in its Case which Incorporates a Battery Charger.

Power Supply

The first prototypes used a repackaged version of the power supply used on military 18-mm image intensifier tubes, and was the second most expensive component in the device. Early in the design phase work on a simplified, lower cost power supply had been initiated. This supply, using a single oscillator instead of the usual two, and working from a 6 volt battery rather than a 2.7 volt one, has somewhat better performance than could have been originally predicted. The power supply is rated for operation only from -18°C (0°F) to 48°C (120°F), but there appears to be no reason why it couldn't be compensated for a wider temperature range at little additional cost.*

Its main performance compromise is that it provides a less flat output luminance in the automatic brightness control region. The tube output has a "hump" in the input region between 10^{-3} and 1 footcandle cathode illuminance. This hump is only typically 3 db and while easily measured instrumentally, is difficult or impossible to detect by eye.

A somewhat unexpected bonus from the power supply development was the fact that this supply occupies about 30% less volume than the repackaged military version.

Cost

The earliest production models of the night vision aid are selling at about one-half the price of industrial-type instruments of similar performance. This is still considerably above our original goal, but it appears that even at the present price a sufficient market exists to support the level of production necessary to establish a learning curve for the product which will lead to further price reductions. At present, not enough information is available to make it possible to forecast the eventual price of the instrument, although an additional factor of two decrease is not an unreasonable forecast.

Conclusion

The night vision aid described above shows promise of being of considerable help to persons suffering from night blindness due to retinal degenerations. Efforts are continuing to reduce the cost of the instrument without serious compomise to present levels of performance.

Acknowledgments

The author would like to gratefully acknowledge the assistance, both administrative and technical, of the following persons: Dr. Steward S. Flaschen, Dr. Eli Arnoff, Messrs. Michael Toohig, William Mims, Alan Hoover, and John W. Smith, all of ITT: Mr. Paul Lighty, and Dr. Douglas Sinclair, independent consultants; Messrs. Malcolm Brookes, Douglas Spranger and Paul Mulhauser of HF/ID Inc.

References

1. Coursey, T.P., McGowan, D.L., Apple, L.E.: "Night Viewing Goggles for Night-Blind Travelers." Bulletin of Prosthetics Research, pp 191-194, 1972.

2. Berson, E.L., Mehaffey, L., Rabin, A.R.: "A Night Vision Device as an Aid for Patients with Retinitis Pigmentosa." Archives of Ophthalmology 90:112-116, Aug. 1973.

3. Berson, E.L., Rabin, A.R., Mehaffey, L.; "Advances in Night Vision Technology," Archives of Opthalmology, 90:427-431, Dec. 1973.

4. Berson, E.L., Mehaffey, L., Rabin, A.R., "A Night Vision Pocketscope for Patients with Retinitis Pigmentosa," Archives of Opthalmology, 91: 495-500, June 1974.

5. Burbo, J.H., "The First Five Generations of Night Vision Devices," Proceedings of Southeastcon 74, pp 330-332, IEEE, 1974.

*A number of patents on this power supply have been applied for by Mr. Alan Hoover.

PHOTOELECTRON MICROSCOPY OF BIOLOGICAL SURFACES. EXCITATION SOURCE BRIGHTNESS REQUIREMENTS

Rudy J. Dam and O. Hayes Griffith

Institute of Molecular Biology and Department of Chemistry,
University of Oregon, Eugene, Oregon 97403

Abstract

Photoelectron microscopy is a surface technique which provides topographical information using the photoelectric effect as a basis for contrast. Progress in the biological applications of this technique is briefly reviewed. Due to relatively low quantum yields, photoemission from biological samples is weak and an image intensifier is used in order to visualize and record the photoelectron image. Currently the limiting magnification is determined by UV power incident on the sample. Power requirements for high-magnification imaging are calculated in terms of microscope, sample, and image intensifier parameters. To approach 40 Å resolution, an instrument magnification of 12,000-50,000 is required along with a UV intensity of 0.01 to 10 Watts/cm^2 depending on the wavelength and sample. For a tightly focused laser source the total power requirement is 1 mWatt or less.

Introduction

Determining the topography of biological surfaces is a challenging problem because of the enormous microheterogeneity in the various proteins, lipids and saccharides present. The photoelectric effect has the potential of mapping the distributions of specific cell surface components without interference from the cellular contents. The basic idea of photoelectron microscopy (PEM) is illustrated in Figure 1. The sample is placed in a vacuum system and subjected to UV light. As the UV source is scanned to shorter wavelengths the surface molecules with the lowest ionization potentials will begin to photoeject electrons which are then accelerated, passed through a series of electron lenses and imaged onto a phosphor screen. This is a very different approach from fluorescence microscopy or transmission or scanning electron microscopy, as shown in Figure 2. There is no electron gun in PEM, the sample is the source of electrons. Indeed, it is a very weak electron source, requiring the use of image intensifiers at even moderate magnifications.

The origin of photoelectron microscopy dates back to the early days of electron microscopy when emission microscopes were constructed to examine hot filaments for use in early transmission microscopes, oscilliscopes and television tubes. Biological surface studies are a relatively new development. The first photoelectron images of mammalian [1] and plant [2] samples are shown in Figures 3 and 4. These are very preliminary low magnification images. Several additional reports examine specific aspects of PEM applications to organic and biological surfaces [3-10]. Metallurgical applications are reviewed elsewhere [11]. Our aim here is to provide a brief overview of the technique and to discuss in detail the UV power requirements for high resolution photoelectron microscopy experiments.

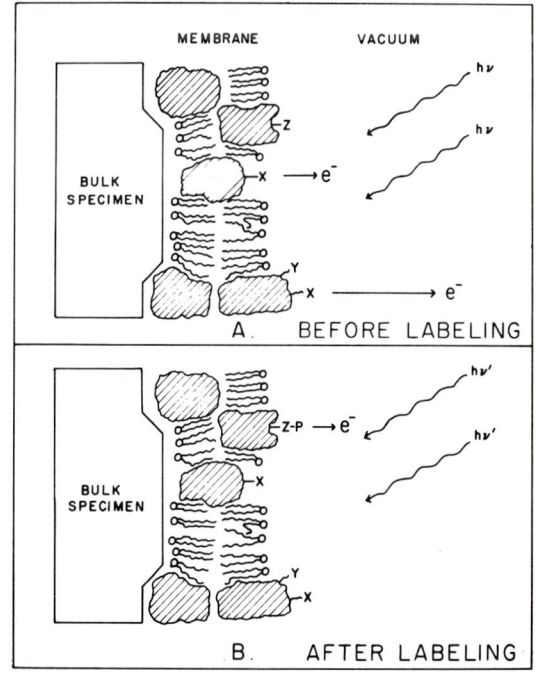

Figure 1. Photoionization of electrons from a hypothetical biological surface. The top diagram illustrates intrinsic photoionization from certain functional groups (X) on the surface. In the bottom diagram, the site Z has been labeled with a photoelectron label P and the energy of the incident light has been lowered from hν to hν', below the ionization threshold of X. Photoelectrons then originate predominantly from sites Z-P. From reference 1.

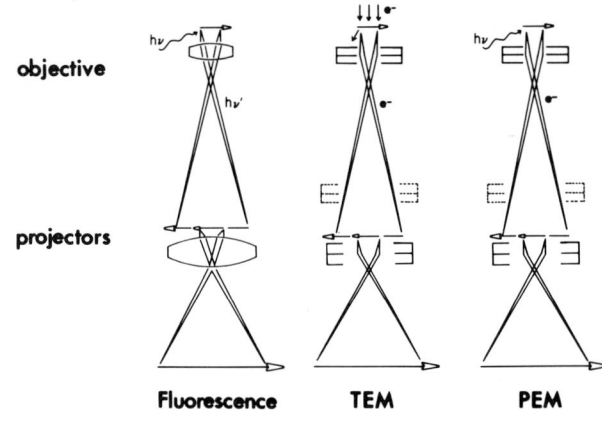

Figure 2. The techniques of fluorescence microscopy, TEM and PEM compared. PEM shares with fluorescence the use of incident exciting light, and with TEM the advantages of electron image formation. From reference 2.

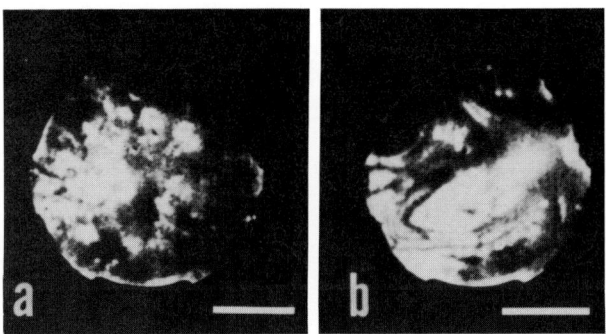

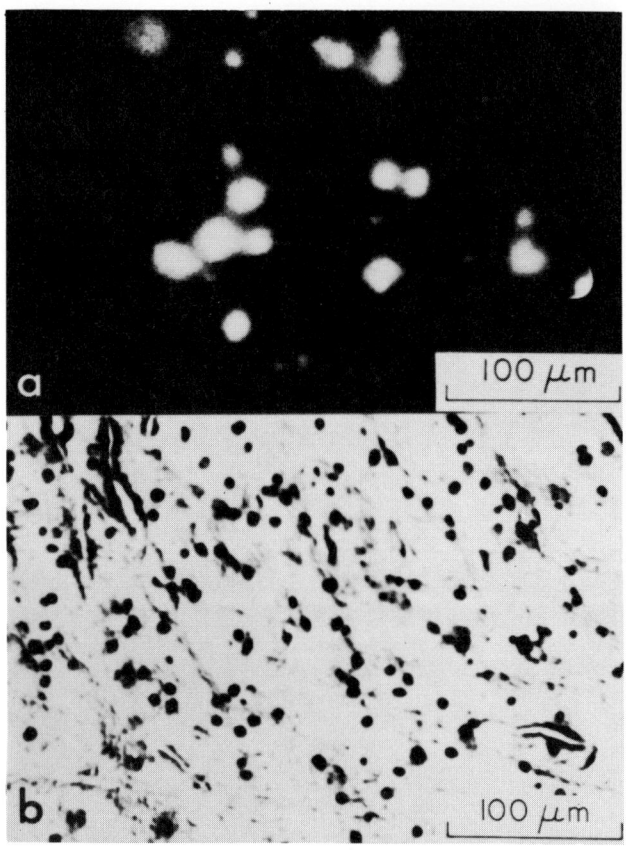

Figure 3. Photoelectron micrographs of sectioned rat epididymis. (a) unfixed unstained tissue section; (b) glutaraldehyde-fixed tissue section. The bar equals 500 μ in (a), 350 μ in (b). Light source: hydrogen lamp (without monochromator); image intensifier: 2.54 cm diameter microchannel plate. From reference 1.

Figure 4. (a) Photoelectron micrograph of a freeze-cleaved chloroplast preparation. Light source: hydrogen lamp, with monochromator set for 200 $\pm$ 10 nm. Image intensifier: 4 cm, three-stage electrostatic. (b) Reflected light micrograph of the same sample (but not the same field of view). The chloroplasts are visible as small black spots against a bright field. From reference 2.

Experimental Apparatus

All data discussed here have been obtained on the prototype photoelectron microscope (a) of Figure 5. This is an oil-free stainless steel ultrahigh vacuum chamber with Varian Conflat copper-sealing flanges and an ion pump to minimize contamination of the sample surface. The two electron lenses are of the electrostatic unipotential type and were designed for very low magnifications (x10 - 200) so that very faint images can be observed, even with a conventional hydrogen discharge lamp-monochromator combination. In Figure 5(a) the monochromatic light is reflected from a magnesium fluoride coated aluminum mirror, through the objective lens, and onto the sample. The light arrives very nearly normal to the sample surface, which is useful in studying the effects of polarization, substrate reflection and optical interference. The sample is at -10kV so that the photoejected electrons are accelerated toward the anode, focused, and passed through a small hole in the mirror, projector lens and onto a microchannel plate image intensifier and phosphor screen. Subsequently the microchannel plate has been replaced by an external 40 mm 3-stage Varo Inc. electrostatic image intensifier coupled to an aluminized phosphor-coated fiber optics window. The aluminized layer

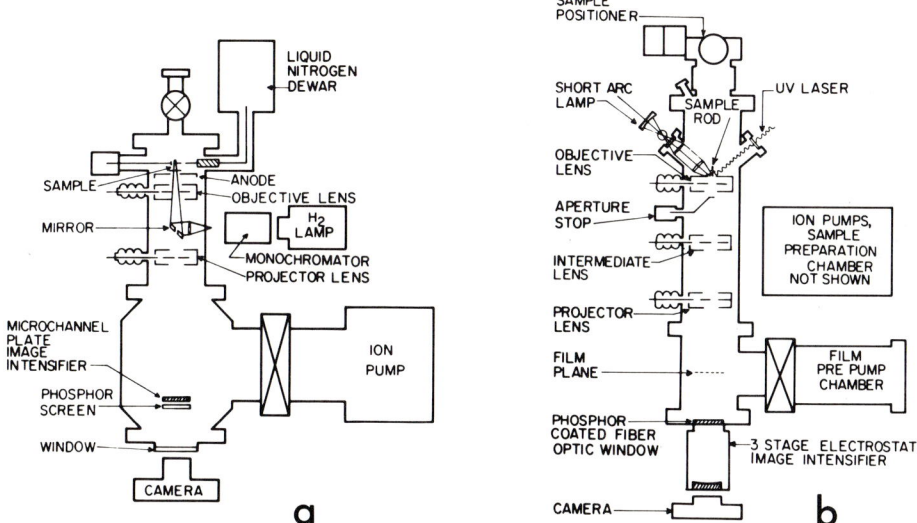

Figure 5. Schematic diagrams of the photoelectron microscope. (a) Prototype low-magnification instrument. (b) High-magnification instrument nearing completion. The major design improvements include a more efficient UV optical geometry with provision for laser excitation, addition of an intermediate electron lens, aperture stop, provision for an internal camera system, and ultrahigh vacuum sample preparation chamber.

reduces the stray UV light reaching the phosphor-image intensifier system.

Figure 5(b) shows the essential features of a second ultrahigh vacuum PEM currently nearing completion at the University of Oregon. The sample may be cooled to 77°K in order to study frozen samples, as in the prototype microscope. The design of the new instrument emphasizes high light intensity from either a conventional UV source (12) or a laser (13), and higher magnification (initially x10,000). Much higher magnifications can be obtained simply by increasing the spacings between the lenses or adding an additional projector lens. The design trade off eliminates the monochromator which is no longer needed after the optimal excitation wavelengths have been determined.

Resolution Factors

There are two resolution factors to consider in PEM studies of surfaces: lateral resolution and depth resolution. The lateral or point-to-point resolution in the plane of the sample is estimated to be 25-40 Å with current electron optics technology (1). However, this has not been tested experimentally since the present limitation is the light source-image intensifier combination. It is for this reason that we consider the light source requirements in detail below. Lateral resolution of 40 Å is sufficient to map the distribution of many protein complexes in membranes but it is not impressive by electron microscopy standards. The advantages of PEM are found in the depth resolution and contrast inherent in the photoelectric effect.

The second and equally important resolution factor is depth resolution. For example, a depth resolution of 100 Å would permit the mapping of a cell surface against the background of the cytoplasm whereas a higher resolution is needed to image the exposed half of the 80-100 Å thick membrane without interference from the inner half of the membrane. More quantitatively, exponential curve fitting of image brightness data vs. sample thickness defines a characteristic depth resolution factor, d_o, from which approximately 60% (i.e., $1-e^{-1}$) of the electrons originate. For the model compound phthalocyanine, $d_o = 15 \text{ Å} \pm 5 \text{ Å}$ (4,14). This is perhaps the highest known depth resolution factor of any microscopic technique. It is a direct result of the very low kinetic energy and hence short escape depth of the photoelectrons.

Contrast

Every molecule has a characteristic set of ionization potentials that contribute to a photoelectron quantum yield curve. There is very little literature on the electron

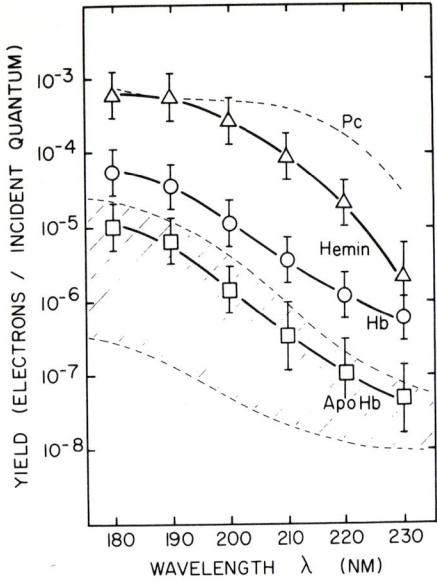
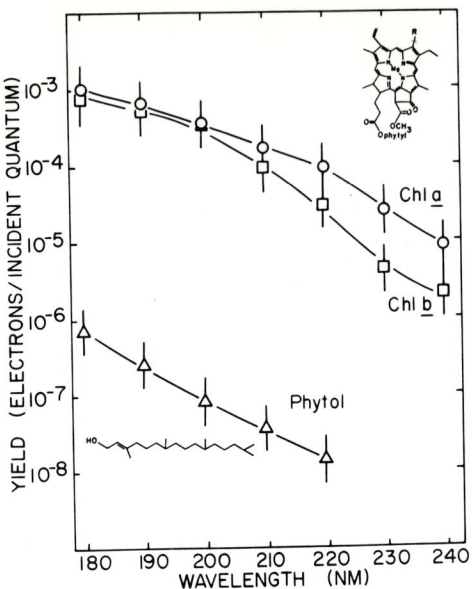

Figure 6. Photoelectron quantum yield curves for hemin, hemoglobin and apohemoblobin (ApoHb). The dashed curve (P_c) is the yield of phthalocyanine reported by Schechtman (15). The shaded band contains the quantum yield data for the amino acids. From reference 7.

Figure 7. Photoelectron quantum yield curves for chlorophyll a (R = CH_3, circles), chlorophyll b (R = CHO, squares) and phytol (triangles). From reference 8.

quantum yields of biological macromolecules. Some representative data are shown in Figures 6 and 7. The photoelectron quantum yields of 19 amino acids fall within the shaded area of Figure 6 (only the aromatic amino acids L-tryptophan and L-tyrosine rise slightly above this band at short wavelengths). Since proteins are composed of amino acids and the yields appear to be additive, the curves for all proteins without prosthetic groups should lie within this band. Apohemoglobin does fall within this band as predicted. Hemin (ferriheme chloride) has a photoelectron quantum yield curve two orders of magnitude larger. The curve for the intact hemoglobin molecule (apohemoglobin + heme) is intermediate and can be estimated from the previous two quantum yield curves assuming a simple dilution model (7). It may prove possible to map the positions for the heme proteins such as the cytochromes using the heme as an intrinsic photoelectron label.

Figure 7 shows the quantum yield curves of the chlorophylls a and b. These curves lie three orders of magnitude above that of the long phytol tail showing that the photoelectric effect is due almost entirely to the porphyrin head group. The quantum yield curves of the chlorophylls are clearly much greater than those of the protein and lipids, so that at high resolution photoelectron microscopy will be useful in mapping chlorophyll distributions in photosynthetic membranes. Chlorophyll is not readily visualized by conventional electron microscopy because the elemental composition and hence electron scattering does not differ greatly from the naturally occurring cell surface components. There are many other photoemissive molecules including phthalocyanine, acridine orange and a carcinogen, benzo(a)pyrine. Tagging antibodies with a photoemissive molecule should permit immunophotoelectric experiments, complimentary to immunofluorescence studies of cell surfaces (1,9).

UV Power Required to Attain 40 Å Resolution

In this section we discuss the UV power requirement problem. A general formula will be derived relating the required UV intensity to a number of microscope and sample variables. Using the best available estimates for their values, we calculate the UV power required to obtain suitable image quality with various sample materials and magnifications. Since the light intensity needed increases as the square of the magnification, attaining 40 Å resolution will require the intensity to increase by a factor of $(50,000/100)^2 = 2.5 \times 10^5$ over the x100 prototype instrument. We have roughly estimated the UV intensity in the prototype PEM to be 10^{-6}–10^{-5} Watt/cm^2 at 200 $\pm$ 10 nm.

This predicts that the intensity required for high resolution would lie in the range 0.2 to 2 Watts/cm^2. We examine the power requirements more quantitatively in the calculations below.

Two conditions must be satisfied in order for the PEM to yield acceptable images at high (x1000 to 100,000) magnifications. The first condition is the presence of sufficient contrast between details of the sample and background to permit image formation. A second condition, related to the quantum statistical nature of imaging with electrons, is the presence of a sufficient number of events (electrons emitted from the sample) per resolution area* during an entire exposure to avoid false detail due to statistical fluctuations in the emission. The first condition depends on a difference in quantum yields between sample and substrate coupled with an optical <u>power</u> requirement, since the rate of build up of photographic exposure differences in sample and background areas of the image depends on incident optical power on the sample compared to the "noise" power artificially created in the microscope within the background areas. The second requirement is an <u>energy</u> requirement, since the statistical fluctuations depend on the total number of incident UV photons striking a resolution area during a complete exposure. We will show that if the first requirement is satisfied, the second can be met with reasonable exposure durations. In the following calculations, we neglect any sample damage caused by heating or photochemistry.

Consider an experiment in which we wish to differentiate between two adjacent surfaces, the "sample" and the "substrate", in the PEM. Let the incident UV flux density be P photons cm^{-2} sec^{-1} of wavelength λ. We define quantum yields (electrons emitted/incident photon) of the sample and substrate to be $Y_s(\lambda)$ and $Y_b(\lambda)$ respectively. The photoelectrons are accelerated to 25 kV and travel down the length of the PEM. They strike a phosphor screen, typically producing 200 photons of green light per incident electron. This light passes through a fiber optic window and is amplified by the three-stage image intensifier with a gain of $\sim 5 \times 10^4$. The resulting image brightness can be either photographed or measured quantitatively using a photomultiplier tube (see Figure 5).

To characterize the experiment completely will require the following additional definitions,

$N_s \equiv$ sample electron emission density, electrons cm^{-2}sec^{-1} leaving sample

$N_b \equiv$ substrate electron emission density

$\gamma \equiv$ photoelectron transport efficiency = fraction of photoelectrons that reach the phosphor screen

$G \equiv$ total gain of phosphor-image intensifier system = (no. of photons leaving output stage of intensifier)/(no. of photoelectrons reaching first phosphor)

$B_s \equiv$ final image flux density of sample in photons cm^{-2}sec^{-1}

$B_b \equiv$ final image flux density of substrate

$B_v \equiv$ image flux density due to intrinsic image intensifier noise (assumed constant)

$M \equiv$ linear magnification of microscope

The imaging system is schematically shown in Figure 8. From this point on we drop the explicit wavelength dependence in our expressions.

Since $N_s = PY_s$ and $N_b = PY_b$, we have

$$B_s = \frac{N_s \gamma G}{M^2} = \frac{PY_s \gamma G}{M^2} \text{ photons cm}^{-2}\text{sec}^{-1} \qquad (1)$$

$$B_b = \frac{N_b \gamma G}{M^2} = \frac{PY_b \gamma G}{M^2} \text{ photons cm}^{-2}\text{sec}^{-1} \qquad (2)$$

---

* "Resolution area" is the area associated with a resolved point image. It is given approximately by r^2 if the lateral point-to-point resolution is r.

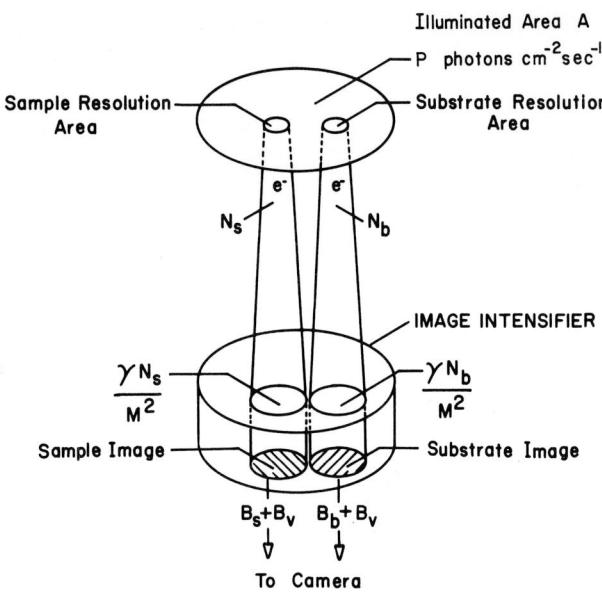

Figure 8. Schematic diagram of the photoelectron microscope-image intensifier system used in the power requirement calculations.

The actual brightness of the sample and substrate images will be $(B_s + B_v)$ and $(B_b + B_v)$ respectively. In order to distinguish between them (e.g., on a photographic plate) the following must be met:

$$(B_s + B_v) \geq K(B_b + B_v)$$

where K is an empirically determined constant chosen to provide adequate contrast. We can rewrite this equation in the form

$$B_s \geq KB_b + (K-1)B_v \qquad (3)$$

Substituting Eqs. (1) and (2) into (3) and rearranging,

$$\frac{P\gamma G}{M^2}\left(Y_s - KY_b - (K-1)B_v\left(\frac{M^2}{P\gamma G}\right)\right) \geq 0 \qquad (4)$$

The first factor is always ≥ 0. Therefore in order for the inequality to hold we must have

$$Y_s - KY_b \geq (K-1)B_v\left(\frac{M^2}{P\gamma G}\right) \qquad (5)$$

Solving for P gives the final result:

$$P \geq \frac{(K-1)B_v M^2}{(Y_s - KY_b)\gamma G} \text{ photons cm}^{-2}\text{sec}^{-1} \qquad (6)$$

We note that when $Y_s \to KY_b$ the intensity requirement becomes arbitrarily large. This simply reflects the fact that no contrast is possible at <u>any</u> power level when the quantum yield of the sample becomes too small, i.e., on the order of KY_b.

Eq. (6) can be manipulated to eliminate the parameters γ and G, so that P depends only on the instrument parameters M, B_v, and K, and the parameters Y_s and Y_b, as follows. We imagine the microscope operating at a given magnification M_0 with no input light on the sample. In this condition the intensifier optical output is B_v, due to noise. We can attribute this B_v to a fictitious electron emission density at the sample, N_v^o electrons cm^{-2}sec^{-1}, in the absence of light. The relation between B_v and N_v^o is therefore

$$B_v = \frac{\gamma G N_v^o}{M_o^2} \qquad (7)$$

If we can determine N_v^o, we can insert (7) into (6) to obtain the value of P at any M.

$$P \geq \frac{(K-1)N_v^o}{Y_s - KY_b}\left(\frac{M}{M_o}\right)^2 \qquad (8)$$

We note that for B_v to be independent of magnification in this model we must have N_v^o varying with M_o such that

$$\left(\frac{N_v^o}{M_o^2}\right) = \text{const.} \equiv N_v' \qquad (9)$$

N_v' is thus a more useful "noise constant" for use in Eq. (8) which becomes

$$P \geq \frac{(K-1)N_v' M^2}{Y_s - KY_b} \qquad (10)$$

By measuring values of B_s for the case $B_s \gg B_v$ with a homogeneous sample and various neutral density filters between the UV source and the sample, we have determined that the imaging system is linear over a three decade range [4]. Then by noting the degree of film darkening corresponding to various input levels when B_s is recorded photographically, we have determined that a one-decade change in B_s gives an appreciable change in film density. Thus K = 10 appears to be a reasonable choice for good image contrast in photomicrography; it may well be an overestimate.

The quantity N'_v was measured as follows. We attached a phototube to the three stage image intensifier and irradiated a homogeneous sample of dye molecules (copper phthalocyanine, with a moderately high Y_S) so as to produce an output brightness $B_S + B_V \approx B_S$. A magnification M_O of 60 was used. The phototube output current I_S was measured as well as the current between the sample and ground in the PEM. The PEM current was 6×10^{-12} amperes and the total sample area was 0.12 cm^2; thus the actual electron emission density at the sample was

$$N_S^O = 3.13 \times 10^8 \text{ electrons cm}^{-2}\text{sec}^{-1} \tag{11}$$

With the UV light off the phototube current I_v was measured, corresponding to the PEM optical noise. Since the system is linear,

$$\frac{I_v}{I_s} = \frac{B_v}{B_s} = \frac{\gamma G N_v^O M_O^2}{\gamma G N_s^O M_O^2} = \frac{N_v^O}{N_s^O} \tag{12}$$

In our experiment $I_v/I_s = 1.28 \times 10^{-2}$, hence $N_v^O = 4.0 \times 10^6$ electrons cm^{-2}sec^{-1} from Eqs. (11) and (12). Finally, since $M_O = 60$ in this experiment, we obtain from Eq. (9)

$$N'_v = \frac{4 \times 10^6}{(60)^2} = 1.11 \times 10^3 \text{ electrons cm}^{-2}\text{sec}^{-1} \tag{13}$$

We are now able to calculate the minimum photon flux using Eq. (10). To describe the input optical quantities in conventional physical units, we note that each photon carries an energy $h\nu = hc/\lambda = 1.986 \times 10^{-25} \times \lambda^{-1}$ joules, where λ is in meters. If we further assume an illuminated area A on the sample, the input power in Watts at the wavelength λ is

$$W \geq \frac{hcA(K-1)N'_v M^2}{(Y_S - KY_b)\lambda} = \frac{2.0 \times 10^{-21} AM^2}{(Y_S - 10Y_b)\lambda} \tag{14}$$

with A measured in cm^2 and λ in meters.

Results of the UV Power Requirements Calculation

We have used Eq. (14) along with results of quantum yield measurements for samples and substrate to calculate estimates of the minimum W for an illuminated area of 0.05 cm^2 (corresponding to an image of a 2.5 mm diameter arc source) and an illuminated area of 10^{-6} cm^2 (attainable with lasers). Table 1 presents the quantum yield values for a Formvar substrate and for two samples, phthalocyanine and poly-L-tryptophan. Entries in Tables 2-4 are milliwatts of average power that will produce minimum acceptable photoelectron images of the samples listed, according to Eq. (14).

Table 1. The absolute electron quantum yield per incident photon for Formvar (Y_b), phthalocyanine, and poly-L-tryptophan as a function of wavelength λ.

λ(nm)	$Y_b(\lambda)$	$Y_s(\lambda)$, phthalocyanine	$Y_s(\lambda)$, poly-L-tryptophan
230	$\ll 10^{-7}$	3.8×10^{-5}	2×10^{-7}
210	$\ll 10^{-7}$	4.0×10^{-4}	2×10^{-6}
190	1.5×10^{-6}	5.2×10^{-4}	6×10^{-5}

The intensity (photons cm^{-2}sec^{-1}) available at the sample is a factor of $.05/10^{-6} = 5 \times 10^4$ higher for a laser than for an arc source of comparable power (the actual ratio may be even larger depending on the efficiency of the optical system used to image the arc). Laser sources in the 180-220 nm region are very promising for this application since the average power required in a tightly focused laser system is quite modest. Consider for example the power requirements in Table 2 for metal-free phthalocyanine. This compound is a very strong photoemitter at all wavelengths shorter than 230 nm and it is used as a standard in our quantum yield measurements. It does not represent the typical case we expect to encounter. Nevertheless, the calculations predict that only $10^{-4} - 10^{-3}$ milliwatts of incident light (per 100 μ^2 sample area) would be required to observe the image of a cluster of phthalocyanine molecules at x100,000. The poly-amino acids of Tables 3 and 4 require higher powers, but still on the order of or less than one milliwatt.

Table 2. Milliwatts of incident light required to view metal-free phthalocyanine using an arc source (A=.05 cm^2) and a laser (A=10^{-6}cm^2).

λ(nm)	$Y_s - 10Y_b$	A, cm^2	\multicolumn{4}{c}{Instrument Magnification, M}			
			100,000	50,000	10,000	1,000
230	3.8x10^{-5}	.05	114	29	1.14	11x10^{-3}
210	4.0x10^{-4}	.05	12	3	0.12	1.2x10^{-3}
190	5.1x10^{-4}	.05	10	2.6	0.10	1.0x10^{-3}
230	3.8x10^{-5}	10^{-6}	23x10^{-4}	57x10^{-5}	23x10^{-6}	23x10^{-8}
210	4.0x10^{-4}	10^{-6}	2.4x10^{-4}	6.0x10^{-5}	2.4x10^{-6}	2.4x10^{-8}
190	5.1x10^{-4}	10^{-6}	2.1x10^{-4}	5.2x10^{-5}	2.1x10^{-6}	2.1x10^{-8}

Table 3. Milliwatts of incident light required to view poly-L-tryptophan.

λ	$Y_s - 10Y_b$	A, cm^2	\multicolumn{4}{c}{Instrument Magnification, M}			
			100,000	50,000	10,000	1,000
230	2x10^{-7}	.05	2.2x10^4	5.4x10^3	2.2x10^2	2.2
210	2x10^{-6}	.05	2.4x10^3	6.0x10^2	24	.24
190	4.5x10^{-5}	.05	1.2x10^2	29	1.2	1.2x10^{-2}
230	2x10^{-7}	10^{-6}	0.43	0.11	4.3x10^{-3}	4.3x10^{-5}
210	2x10^{-6}	10^{-6}	4.8x10^{-2}	1.2x10^{-2}	4.8x10^{-4}	4.8x10^{-6}
190	4.5x10^{-5}	10^{-6}	2.3x10^{-3}	5.9x10^{-4}	2.3x10^{-5}	2.3x10^{-7}

Table 4. Milliwatts of incident light required to view poly-L-arginine HCl. The value of $(Y_s - 10Y_b)$ is quite uncertain and was here taken to be 0.1x10^{-5}.

λ	$Y_s - 10Y_b$	A, cm^2	\multicolumn{4}{c}{Instrument Magnification, M}			
			100,000	50,000	10,000	1,000
180	1x10^{-6}	.05	5.6x10^3	1.4x10^3	56	0.56
180	1x10^{-6}	10^{-6}	0.11	2.8x10^{-2}	1.1x10^{-3}	1.1x10^{-5}

The absence of entries in Table 4 for wavelengths other than 180 nm reflects the fact that the quantum yield of this polymer is on the order of or less than 10 Y_b for a Formvar substrate. Thus, at wavelengths longer than 180 nm, little or no contrast would be possible regardless of input power. This is fortunate. The preliminary data suggest that many polypeptides and other possible surface components will contribute minimal background signals, which means higher contrast in photoelectron labeling experiments. (The analogy in fluorescence microscopy is a low intrinsic fluorescence.)

Having established the UV power levels required for high-magnification work, we now consider the relationship between magnification and resolution.

The Dependence of Resolution on Magnification

The limiting resolution of an optical system whose separate components have resolution limits $r_1, r_2, \ldots$ is given by

$$r_{sys} = (r_1^2 + r_2^2 + \ldots)^{\frac{1}{2}} \tag{15}$$

Consider a two-component system consisting of the PEM and image intensifier. For the purposes of this calculation we take the lateral resolution of the PEM as $r_1 = 40$ Å. The resolving power of the three-stage image intensifier is 28 lp/mm corresponding to 36 μ at the intensifier output stage. To refer this resolution limit back to the specimen plane in Å we divide by the instrument magnification M (the intensifier magnification is about 1), so that $r_2 = 3.6 \times 10^5/M$ (Å). Thus we obtain for this system

$$r_{sys} = (1600 + 1.3 \times 10^{11}/M^2)^{\frac{1}{2}} \text{ Å} \tag{16}$$

The system resolution is limited mainly by the image intensifier at low magnifications (M < 5,000) and by the microscope at high magnifications (M > 20,000). At intermediate M both components contribute significantly to r_{sys}.

Further magnification is often required to visualize r_{sys} on the final micrograph. In most cases the image on the intensifier output will be photographed through a conventional optical system to produce the final negative or print. If m is the photographic magnification introduced by this last step, the image separation between two point objects at the system limit is $r_{sys}Mm$. In order to be visible to the eye this separation must be of order 100 μ = 10^6 Å at a standard viewing distance of 25 cm under optimum conditions. This figure is often increased to 250 μ = 1/4 mm for ease of visibility (16).

Adopting the latter value of 250 μ, we calculate first the minimum M required to achieve a given r_{sys} using Eq. (16) and the subsequent photographic magnification m required to achieve 250 μ separation on the final print. The same final result could also be achieved without photographic enlargement by increasing M. Thus a range of possible M values exists which will provide the required image separation for a given r_{sys}. The following table lists the results of these calculations. The smallest M value in each range corresponds to the solution of Eq. (16) and requires the largest m in order to realize the final print magnification Mm.

Table 5. Magnification required to achieve a given resolution r_{sys}

r_{sys}, Å	range of M	range of m	mM
50	12,000-50,000	4.2-1	50,000
100	3,600-25,000	7-1	25,000
1000	360-2500	7-1	2,500

These calculations show that the limiting resolution of the microscope can be approached using instrument magnifications in the range 12,000 - 50,000 and the incident power levels shown in Tables 2-4.

Statistics of the Photoelectron Image

To evaluate the effect of statistical fluctuations in the electron emission, we consider the number of electrons emitted from a resolution area element r_{sys}^2 during a reasonable exposure time with a given UV power input $P = \beta P_{min} \geq P_{min}$. Recall that

$$P_{min} = \frac{(K-1)N_v'M^2}{Y_s - KY_b} \text{ photons cm}^{-2}\text{sec}^{-1} \tag{10}$$

Consider an incident power βP_{min}. The sample emission density with this power input is $N_s = \beta P_{min} Y_s$ electrons cm^{-2} sec^{-1}. We follow the electrons through the PEM, image intensifier, and camera system. Assuming no losses in the electron optics, $\gamma = 1$ and the photon flux density at the output stage of the image intensifier is

$$F_{int} = B_s + B_v \approx GN_s/M^2 = \frac{\beta G(K-1)N_v'}{1 - KY_b/Y_s} \text{ photons sec}^{-1}\text{cm}^{-2} \tag{17}$$

The noise contribution term $B_v = GN_v'$ has been neglected since it will be smaller than B_s by at least a factor of $\beta(K-1) \approx 10$. If this image is photographed with an optical system of efficiency ε, the flux density at the film plane will be

$$F_{film} = F_{int}\varepsilon = \frac{\beta\varepsilon G(K-1)N_v'}{1 - KY_b/Y_s} \tag{18}$$

Optical efficiency here is defined as the photon flux received at the film plane divided by the flux at the output stage of the image intensifier. It can be estimated from the optical parameters of the camera system and the magnification m. If the source radiates uniformly into 2π steradians, ϵ is of order 10^{-2} for a lens system of focal ratio F/1.2 working near unit magnification.

In order to calculate the exposure time it is convenient to convert photons $sec^{-1} cm^{-2}$ to foot-candles. For $\lambda = 550$ nm light the conversion factor is 1 photon $cm^{-2} sec^{-1}$ = 2.26×10^{-13} ft-c (see, e.g., reference 17). Inserting the previously calculated parameters $G = 5 \times 10^6$ and $N_v' = 1.11 \times 10^3$ into Eq. (18) we obtain

$$F'_{film} = 1.25 \times 10^{-3} \frac{\beta \epsilon (K - 1)}{1 - KY_b/Y_s} \quad \text{ft-c} \tag{19}$$

The exposure time required to obtain an acceptable film density is given by [16]

$$t \approx \frac{2.25}{F'_{film} \cdot (ASA)} \quad \text{sec} \tag{20}$$

where (ASA) is the film speed. Inserting Eq. (19) into Eq. (20),

$$t = 1.8 \times 10^3 \frac{1 - KY_b/Y_s}{\beta \epsilon (K-1)(ASA)} \tag{21}$$

As an example, the exposure time for a high-contrast sample ($Y_s \gg 10 Y_b$) assuming $\epsilon = 5 \times 10^{-3}$, $K = 10$ and ASA = 400, is $t = 100$ sec for $\beta = 1$ ($P = P_{min}$) and $t = 10$ sec for $\beta = 10$ ($P = 10 P_{min}$). The exposure time can also be decreased by increasing ϵ or the film speed. However, the parameters β, ϵ, and film speed also affect the image quality as shown in the following argument.

If we assume that individual photoelectron events are detectable, the statistical properties of the image are determined by the total number of electrons emitted from a sample resolution area element r^2_{sys} during a typical exposure. In a time t such an area element emits

$$N = N_s t \, r^2_{sys} \quad \text{electrons} \tag{22}$$

Substituting the derived expressions for t and r_{sys} from Eqs. (21) and (16) and setting $N_s = \beta P_{min} Y_s$, we obtain for this system

$$N = 2 \times 10^{-10} \frac{M^2}{\epsilon \cdot (ASA)} [1600 + 1.3 \times 10^{11}/M^2] \quad \text{electrons} \tag{23}$$

Assuming Poisson statistics apply to the electron emission we may define a signal-to-noise ratio $N/\delta N \approx N/\sqrt{N} = \sqrt{N}$. The result is

$$S/N = 1.4 \times 10^{-5} \frac{M}{[\epsilon \cdot (ASA)]^{\frac{1}{2}}} [1600 + 1.3 \times 10^{11}/M^2]^{\frac{1}{2}} \tag{24}$$

Note that increasing ϵ or the film speed in order to decrease the exposure time t adversely affects S/N, since a smaller number of electrons are emitted from a resolution element during the exposure. However, S/N is <u>independent</u> of $\beta = P/P_{min}$ since an increase in power input is exactly balanced by a decrease in exposure time, so that the total number of electrons emitted during the exposure remains constant. Thus increasing the incident power is a desirable method of reducing t without sacrificing S/N. Returning to the above example, the S/N for various values of M are:

M	S/N	($\epsilon = 5 \times 10^{-3}$; K = 10; ASA = 400)
≤ 1,000	4	
10,000	5	
25,000	10	
50,000	20	

The interesting prediction that S/N will increase with magnification results from the behavior of r^2_{sys} with M. As the magnification is increased, the incident power required to overcome intrinsic noise must increase as M^2 according to Eq. (10). At low M, $r^2_{sys} \propto 1/M^2$ which cancels this factor in the expression for N (Eq. (22)) and S/N is roughly constant. At high M $r_{sys} \approx$ constant and $N \propto M^2$ due to the power requirement term; thus $S/N \propto M$.

A S/N of ≈ 5 is required for distinguishing detail at the resolution limit[18], and thus the predicted values for $M \geq 10,000$ are quite acceptable. With these operating parameters, some loss of resolution may occur at low magnifications. This is not likely to be a major drawback, since most high-resolution work will be done at $M \geq 10,000$.

As shown previously, the exposure time t may be reduced from 100 sec at minimum UV power to 10 sec by using $P = 10P_{min}$. In order to bring about a similar change by increasing ε or (ASA), the S/N ratio would decrease by a factor of $\sqrt{10} \approx 3$, which in this case would be unacceptable, especially in the lower range of M. While this sample calculation may not describe exactly a given system, it does point out the compromises involved in obtaining a final image of high quality.

Conclusions

The above calculations specify the conditions needed to proceed with high-resolution photoelectron microscopy of biological surfaces. In order to approach 40 Å lateral resolution a combined instrument and photographic magnification of x50,000 is required. The corresponding minimum UV intensity varies with sample and wavelength, as shown in Tables 2-4. This variation is due to the wavelength dependence of the photoelectron quantum yields, which differs widely among organic and biological samples. Components with higher yields require lower incident intensities to reach a given magnification. For example, the dye phthalocyanine requires between 0.002 and 0.06 Watts/cm^2 for 40 Å resolution imaging against a Formvar substrate. Similar intensities are expected for heme-containing biological samples. A compound like poly-L-tryptophan requires from 0.5 to 10 Watts/cm^2. These intensity values compare favorably with the order of magnitude estimates based on the present limiting magnification and UV intensity of the prototype instrument. The more detailed calculations show explicitly the dependence of incident UV intensity on sample, substrate and microscope parameters.

The total power requirement depends on the type of source used and the UV optical efficiency. In this regard a laser source offers considerable advantages over a conventional arc or discharge lamp. For a tightly focused laser source ($A = 100\ \mu^2$), the optical power needed to image proteins against a dark substrate lies in the 10^{-3} - 0.05 milliwatt range. If the beam were defocused by a factor of 10 in area, the total power needed is still less than or on the order of 1 milliwatt. Finally, regardless of the type of source, statistical considerations show that with reasonable film and camera parameters, acceptable signal-to-noise ratios are obtainable throughout the useful magnification range of the instrument.

Acknowledgments

We are indebted to Drs. Gail Massey and Charles A. Burke and also to George H. Lesch for useful discussions. The photoelectron microscopes are funded by PHS Grant no. CA 11695 from the National Cancer Institute. The light source development is funded by NSF grant no. BMS 74-11824. R.J.D. and O.H.G. wish to acknowledge support from grant no. GM 00715-17 and CDA Grant no. 1 K04 CA23359 from the National Cancer Institute, respectively.

References

1. Griffith, O.H., Lesch, G.H., Rempfer, G.F., Birrell, G.B., Burke, C.A., Schlosser, D.W., Mallon, M.H., Lee, G.B., Stafford, R.G., Jost, P.C. and T.B. Marriott. Photoelectron Microscopy: A New Approach to Mapping Organic and Biological Surfaces. Proc. Nat. Acad. Sci. USA 69, 561 (1972).
2. Dam, R.J., Lesch, G.H., Deamer, D.W. and O.H. Griffith. Photoelectron Microscopy of Biomembranes: Observation of External Photoemission from Spinach Chloroplasts. Proc. Electron Microscopy Soc. Amer. 33, 502 (1975).
3. Birrell, G.B., Burke, C.A., Dehlinger, P. and O.H. Griffith. Contrast in the Photoelectric Effect of Organic and Biochemical Surfaces. A First Step Towards Selective Labeling in Photoelectron Microscopy. Biophys. J. 13, 462 (1973).
4. Burke, C.A., Birrell, G.B., Lesch, G.H. and O.H. Griffith. Depth Resolution in Photoelectron Microscopy of Organic Surfaces. The Photoelectric Effect of Phthalocyanine Thin Films. Photochem. Photobiol. 19, 29 (1974).
5. Dam, R.J., Burke, C.A. and O.H. Griffith. Photoelectron Quantum Yields of the Amino Acids. Biophys. J. 14, 467 (1974).
6. Engel, W. and S. Grund. Photoelectron Emission Microscopy of Biological Specimens. Proc. 8th Intl. Congress on Electron Microscopy, Canberra, (1974); Vol. II, p. 656.
7. Dam, R.J., Kongslie, K.F. and O.H. Griffith. The Photoelectron Quantum Yields of Hemin, Hemoglobin and Apohemoglobin: Possible Applications to Photoelectron Microscopy of Heme Proteins in Biological Membranes. Biophys. J. 14, 933 (1974).
8. Dam, R.J., Kongslie, K.F. and O.H. Griffith. Photoelectron Quantum Yields and Photoelectron Microscopy of Chlorophyll and Chlorophyllin. Photochem. Photobiol. 22, 265 (1975).
9. Grund, S., Engel, W. and P. Teufel. Photoelektronen Emissionsmikroskop und

Immunofluoreszenz. J. Ultrastructure Res. 50, 294 (1975).

10. Dam, R.J., Rempfer, G.F. and O.H. Griffith. Photoelectron Microscopy of Organic Surfaces: The Effect of Substrate Reflectivity. J. Appl. Phys., in press (1976).

11. Wegmann, L. The Photo-Emission Electron Microscope. Its Technique and Applications. J. Microscopy 96 (1), 1 (1972).

12. Engel, W. Emission Microscopy with Different Kinds of Electron Emission. Proc. 6th Intl. Congress Electron Microscopy, Kyoto; 217 (1966).

13. Massey, G.A. Efficient Upconversion of Long-Wavelength UV Light into the 200-235 nm Band. Appl. Phys. Lett. 24, 371 (1974).

14. Pong, W. and J.A. Smith, Photoelectric Emission from Copper Phthalocyanine. J. Appl. Phys. 44, 174 (1973).

15. Schechtman, B.H. Photoemission and Optical Studies of Organic Solids: Phthalocyanines and Porphyrins. Technical Report no. 5207-2 from Stanford Electronics Laboratories, Stanford University, Stanford, California (1968).

16. Loveland, R.P. Photomicrography, Vols. I and II, John Wiley & Sons, New York (1970).

17. Klein, M.V. Optics, John Wiley & Sons, New York (1970).

18. Valentine, R.C. The Response of Photographic Emulsions to Electrons, in Advances in Optical and Electron Microscopy, Vol. I, R. Barer and V.E. Cosslett, eds., Academic Press, New York (1968).

THE APPLICATION OF LOW LIGHT LEVEL VIDEO TECHNIQUES
TO BIOMEDICAL RESEARCH

James A. Dvorak
Laboratory of Parasitic Diseases
National Institute of Allergy and Infectious Diseases
National Institutes of Health, Bethesda, Maryland 20014

William H. Schuette
Biomedical Engineering and Instrumentation Branch
Division of Research Services, NIH

Willard C. Whitehouse
Television Engineering Section
Patient Services Department
Clinical Center, NIH

Abstract

Low-light-level video techniques are applicable to the non-destructive study of biological phenomena at both the organ or macroscopic level and the cellular or microscopic level. At the macroscopic level, a low-light-level video fluorometer has been employed to study transient NADH changes occuring during cortical brain stimulation and ischemic changes occuring during myocardial infarction. At the microscopic level, low-light-level video techniques have been employed to study rapidly occuring transient events in living cells that are photosensitive at light levels normally employed for visualization.

Introduction

Low-light-level video technology, once the almost exclusive domain of defense and aerospace disciplines is becoming increasingly important in other disciplines as well. This report outlines some of the more recent, less obvious applications of low-light-level video techniques to biomedical research. More specifically, it describes two applications of low-light-level video technology that have been studied intensely at the National Institutes of Health.

Video Fluorometry of Intact Organs

The first project to be discussed involves the study of metabolic and electrical activity in the cerebral cortex of the exposed brain. The focus of the study has been to relate oxidative metabolism in the intact brain as indicated by NADH fluorescence to potassium kinetics and, subsequently, both of these functions to direct oxygen consumption. The simultaneous monitoring of potassium kinetics and NADH fluorescence could provide insight into the relationship between ion transport and oxidative metabolism in the intact brain during increased metabolic and electrical activity.

Initially the project centered around the design and construction of a dual beam fluorometer intended for use in the monitoring of reduced nicotinamide adenine dinucleotide (NADH). Reduced NADH has been used extensively as an indicator of oxidative metabolism. Reduced NADH fluoresces with a broad band of 425-475nm light when excited with near ultra violet radiation from 310-370nm, whereas oxidized nicotinamide adenine denucleotide (NAD+) does not fluoresce significantly. It was assumed that the NADH fluorescence signal would be proportional to the rate of oxygen utilization. Thus a linear correlation was sought for the NADH fluorescence and the magnitude of the potassium change obtained as a result of electrical stimulation.

The video fluorometer for this project consists, in essence, of 3 parts (Fig. 1). These are a UV light source, a television camera with a custom made receiver consisting of a prism, filter block and image intensifier, and the video integrator for signal analysis [1]. The UV illuminated tissue is viewed by a low-light-level video camera system with two wedge-shaped prisms arranged in a holder fitted to two separate optical filters. This arrangement enables the presentation of two video images of equal optical path length to the camera lens. The thicker portions of the prisms are positioned over the center of the camera lens in order to prevent peripheral field fall off.

Filters for passing light in the 425-475nm NADH fluorescence band are mounted in front of one prism and the other prism is covered by a narrow-band interference filter with a peak at 529nm. The resultant video display consists of two images of the illuminated exposed brain. One image represents NADH fluorescence and the other represents fluorescene from sodium fluorescein, the reference standard for this study. In practice, the image pair is passed to a one inch video tape recorder for storage and analysis. A typical example of the video display is shown in Figure 2.

Analysis can be either on-line or by replay of videotape. The video signal is processed by an electronic integrator [2] to express differences in specimen density as an analog signal representing proportional changes in integrator voltage. An area-of-interest integrator window is positioned over the video image of the specimen and its size and aspect ratio is operator-determined in raster to restrict the field of interest.

Initial results utilizing this instrumentation indicate that three parameters namely, NADH, O_2 consumption and potassium kinetics are closely interrelated. Thus, the low-light-level video system can be used for topographical studies of redox changes occuring during cortical brain activity in vivo. Transient NADH changes measured by the fluorometric system after electrical stimulation and with epileptiform activity are similar in size and morphology to those described previously (3,4). As all of the data can be recorded on video tape and replayed several times and the integrator window can be positioned over a different area of the specimen for each tape replay this system is capable of measuring changes occuring simultaneously in many different areas of brain and should prove useful in studies of the relative distribution of some metabolic changes that occur on the exposed cortex. The fluorescence mapping technique appears to be more spatially definitive than electrocorticogram mapping and may provide an accurate localization of epileptiform foci in the human cerebral cortex and the differentiation of projected and local discharges. Consequently, this study has immediate relevance for our understanding of epileptic seizures in man.

Recently, the video fluorometer has been utilized in a study of ischemic changes occuring during myocardial infarction. The instrumentation used and parameters measured are similar to those just described. This study has direct relevance in studies of both acute and chronic heart disease in man. Thus, the concept of low-light-level video fluorometry probably has broad applicability in both research and applied clinical medicine.

Low Light Level Video Microscopy

Another application of low-light-level video techniques being pursued at the NIH involves the collection and processing of images of living cells produced by a microscope.

Stating the problem in simplist terms, this requires the recording of rapidly occuring events through a microscope equipped with polarization optics at a final optical magnification in excess of 1000 diameters at high resolution and low-light-level. The use of living material permits kinetic analysis of events; the use of polarization optics permits the application of quantitative techniques; the use of high optical magnification and the need for high resolution are dictated by the size of the object being studied; the use of low-light-level is dictated by the extreme photo-sensitivity both qualitative and quantitative of the living cells being studied.

One example of the successful use of low-light-level video microscopy for the study of a living system involves the interaction of the organism causing malaria with the red blood cell.

Malaria constitutes one of the major maladies affecting mankind today. Although it is known that malaria parasites invade and reproduce within red blood cells the actual process of invasion had not been observed or recorded. This is a particularly crucial sequence of events, for if the penetration of erythrocytes by merozoites could be interrupted, the red blood cell cycle of the parasite would be broken and the malaria infection terminated. The object of our work was to record this interaction in order to describe it phenomenologically and, subsequently, to develop an assay method for studies involving the attempted interdiction of the event at the cellular and subcellular levels. The living red blood cells were inoculated along with cells containing mature parasites into a controlled-environment culture system capable of being maintained at steady-state physiologic conditions (5). The major component of the culture system is a chamber designed to fulfill the following requirements:
1. Attain the highest possible light optical resolution using high numerical aperture objectives and condensers.
2. The final assembly must be strain-free to permit the utilization of polarized light techniques.
3. Maintain a completely closed system to permit the cultivation of human pathogens or other hazardous material.
4. Laminar flow perfusion of nutrients to maintain a uniform and defined environment for the material being cultured.

Figure 3 shows a culture chamber on the stage of a microscope. Figure 4 shows a cross-sectional assembly drawing of the chamber and its relationship to the objective and condenser of the microscope. The image of the red blood cells is projected onto the faceplate of a low-light-level video camera and following processing with a shading generator and image enhancer is recorded on one inch video tape. Selected data is subsequently transfered to 16mm motion picture film with an electron beam recorder.

The results of this study have been published (6). However, a summary of the major findings may provide some appreciation for the combination of biological and electro-optical problems involved in this type of work.

Following the reproductive and developmental phase of the parasite within a red blood cell, the parasite must leave this expended red blood cell and invade another red blood cell in order to complete its life cycle. The rupture of the infected red blood cell is preceded by the coalescence of malaria pigment into a single residual body (Fig. 5) (Scale = 5um). This is followed by vesiculation of the red cell membrane and finally swelling and rupture of the red blood cell and release of the parasites with explosive suddenness.

The penetration of another red blood cell by the liberated parasite occurs only if the anterior end of the parasite containing the paired organelles, contacts the red blood cell (Fig. 6). The initial attachment between the anterior end of the parasite and red blood cell results in a rapid and marked deformation of the red blood cell for a period of 5 to 10 seconds. Deformation occurs irrespective of the site on the red blood cell contacted by the anterior end of the parasite. The parasite remains attached to the red blood cell but does not enter the cell during this initial deformation. The actual interiorization of a parasite requires approximately 10 to 20 seconds for completion. Following this interiorization the red blood cell is again deformed. This second wave of deformation continues intermittently for 10 to 15 minutes, after which the parasite becomes quiescent and the red blood cell resumes its biconcave shape.

The major questions arising from this study concern the actual mechanisms involved in red blood cell deformation and parasite interiorization. Severe red blood cell deformation occurs twice-after initial attachment of the parasite and again after complete interiorization of the parasite. These changes are probably induced by the parasite, but neither the site or sites affected in the red blood cell nor the physical and chemical nature of the initial membrane lesion is known.

Based on the data available to date, a theoretical mode for the penetration of red blood cells by malaria parasites can be devised (Fig. 7). Free parasites come in contact with red blood cells by chance. Properly oriented contact results in the attachment of parasites to red blood cells through an interaction between receptor sites on the anterior end of the parasite and red blood cell surface. A localized lesion may be produced in the red blood cell membrane at the site of attachment. Widespread deformation of the red blood cell could occur as a result of this lesion. After this deformation has subsided, the parasite enters the red blood cell by producing a localized invagination of the red blood cell membrane. Although within the red blood cell, the parasite is still exposed to the external milieu through the orifice of the resulting invagination. The orifice and membrane lesion permit sodium, water, and possibly calcium to enter the red blood cell at a rate in excess of its normal exchange capacity. However, the lesion is too small to permit hemoglobin or other large molecules to leave the red blood cell. The membrane lesion could be isolated from the external milieu by fusion of the red blood cell membrane at the orifice of the invagination. After fusion, the normal intracellular water and electrolyte content of the red blood cell is restored and the cell serves as a host for development of the malaria parasite.

These data could not have been obtained without the use of low-light-level video instrumentation. Subsequent work in which low-light-level video instrumentation was used as an assay tool established that certain aspects of the process of invasion of a red blood cell by a malaria parasite are modulated by genetic factors expressed on the surface of the red blood cell (7). The chemical definition of these factors could result in the ability to produce a vaccine against this disease.

The data just discussed is an example of successful utilization of low-light-level video microscopy. There are many other examples where we have failed. These failures appear to be due to two parameters that are interrelated. That is, collector efficiency or system gain and spatial resolution. The localization of NADH at the subcellular level, for example, would have far reaching implications in biomedical research. This can not be accomplished with presently available low-light-level video instrumentation as there is no way to change these biological systems to improve their performance as photon transmitters. Therefore, it appears that the only hope we have is an improvement in receivers, that is, low-light-level devices. The RCA image isocon tube we are presently using is rated at 200 TV lines with a 25% contrast image and an illuminance of 2×10^{-5} ft-candles on the faceplate. Both of these tubes have been very useful for our work. However, we commonly deal with low contrast extended gray images having a total dynamic range of only 5 - 10% even when operating the low-light-level devices at their maximum rated sensitivity. Consequently, we are unable to record some of the most interesting biomedical phenomena. As end users of low-light-level devices we wish for continued progress in the production of even more sensitive instrumentation for use in biomedical research.

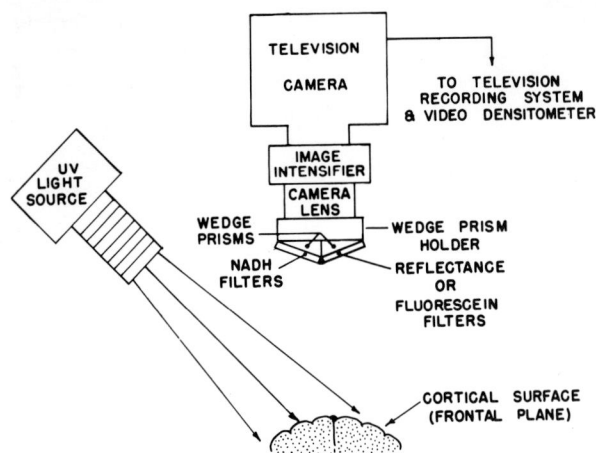

Figure 1. Schematic diagram of low-light-level video fluorometer used for studies of NADH fluorescence of exposed organ surfaces.

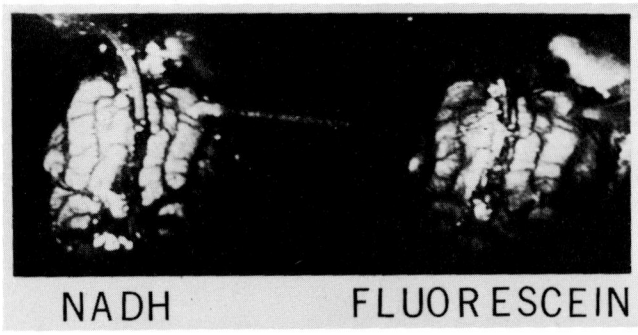

Figure 2. Representative example of NADH fluorescence of exposed brain cortex and fluorescein control.

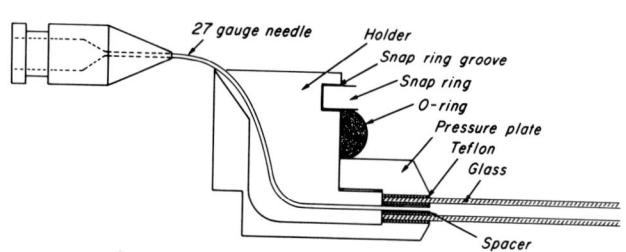

Figure 3. Partial cross-sectional assembly drawing of controlled-environment culture system used for low-light-level video studies of living vertebrate cells. The chamber is designed to fullfill stringent light microscopy requirements while maintaining optimum physiologic conditions.

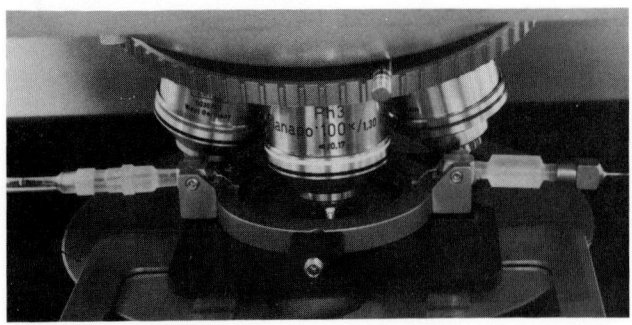

Figure 4. Representative example of the installation of a controlled-environment culture system on the stage of a compound research light microscope.

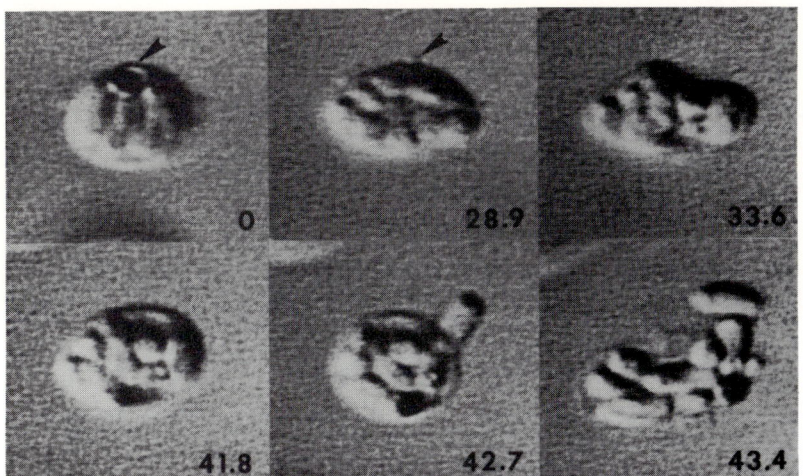

Figure 5. This sequence of photographs depicts the rupture of a red blood cell containing mature parasites. The arrow points out the residual body and the numbers represent the time in minutes from the beginning of the sequence.

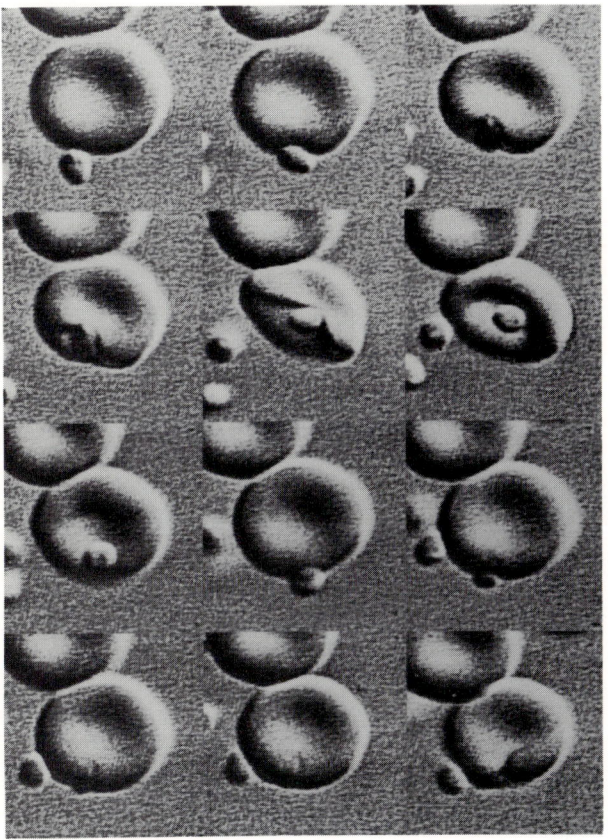

Figure 6. This sequence of photographs depicts from upper left to lower right the invasion of a red blood cell by a malaria parasite. Note the marked deformation of the red blood cell that occurs following properly oriented contact of the malaria parasite with the red blood cell and, subsequently, during the actual process of invasion

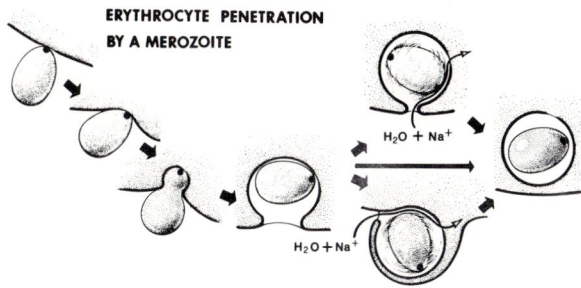

Figure 7. Schematic representation of sequence of events occuring during the penetration of a red blood cell by a malaria parasite.

References

1. Schuette, W. H., Whitehouse, W. C., Lewis, D. V., O'Connor, M. and Van Buren, J. M., "A Television Fluorometer For Monitoring Oxidative Metabolism in Intact Tissue," Medical Instrumentation 8:331, 1974.
2. Dvorak, J. A., Schuette, W. H. and Whitehouse, W. C., "A Simple Video Method For The Quantification of Microscopic Objects," Journal of Microscopy 102:71, 1974.
3. Jobsis, F. F., O'Connor, M. J., Vitale, A. and Vreman, H., "Intracellular Redox Changes in Functioning Cerebral Cortex. I. Metabolic Effects of Epileptiform Activity," Journal of Neurophysiology 34:735, 1971.
4. Rosenthal, M. and Jobsis, F., "Intracellular Redox Changes in Functioning Cerebral Cortex. II. Effects of Direct Cortical Stimulation," Journal of Neurophysiology 34:750, 1971.
5. Dvorak, J. A. and Stotler, W. F., "A Controlled-environment Culture System For High Resolution Light Microscopy," Experimental Cell Research 68:144, 1971.
6. Dvorak, J. A., Miller, L. H., Whitehouse, W. C. and Shiroishi, T., "Invasion of Erythrocoytes by Malaria Merozoites," Science 187:748, 1975.
7. Miller, L. H., Mason, S. J., Dvorak, J. A., McGinniss, M. H. and Rothman, I. K., "Erythrocyte Receptors For (Plasmodium Knowlesi) Malaria: Duffy Blood Group Determinants," Science 189:561, 1975.

Author Index

Aikens, R. S., *Astronomical Applications of Charge Injection Devices,* 65
Antcliffe, Gault A., *A Large Imaging Array CCD Program,* 60
Baker, R., *High-Resolution Low-Light-Level Video Systems for Diagnostic Radiology,* 102
Boyle, J., *Operation of CCDs in the Electron Bombarded Mode,* 10
Brennan, Thomas M., *Unusual Applications of Image Intensification Devices,* 55
Brill, A. B., *Methods and New Approaches to the Calculation of Physiological Parameters by Videodensitometry,* 118
Burbo, James H., *A Night Vision Aid as a Consumer Product,* 137
Caldwell, L., *Operation of CCDs in the Electron Bombarded Mode,* 10
Capp, P., *High-Resolution Low-Light-Level Video Systems for Diagnostic Radiology,* 102
Carruthers, G., *Low Light Level Imaging Devices for the Middle Ultraviolet,* 95
Choisser, John P., *A Photon Counting Array Photometer,* 83
Currie, Douglas G., *A Photon Counting Array Photometer,* 83
Dam, Rudy J., *Photoelectron Microscopy of Biological Surfaces—Excitation Source Brightness Requirements,* 143
Derniak, E. L., *Results of Ratio Temperature Thermography,* 126
Dvorak, James A., *Application of Low Light Level Video Techniques to Biomedical Research,* 155
Dyer, William H., *Unusual Applications of Image Intensification Devices,* 55
Fender, Ferd, *General Application of Microchannel Image Inverters,* 28
Frost, M., *High-Resolution Low-Light-Level Video Systems for Diagnostic Radiology,* 102
Graham, T. P., Jr., *Methods and New Approaches to the Calculation of Physiological Parameters by Videodensitometry,* 118
Griffith, O. Hayes, *Photoelectron Microscopy of Biological Surfaces—Excitation Source Brightness Requirements,* 143
Gur, David, *Isocon Imaging for X-Ray Diagnostics,* 108
Haas, David J., *The Technology behind X-Ray Security Systems,* 44
Hall, James A., *Comparison of TV Imagers for Use in Low-Light-Level Imaging by Electron Beam Scan vs Solid-State Readout,* 14
Hicks, G., *Low Light Level Imaging Devices for the Middle Ultraviolet,* 95
Hoagland, Kenneth A., *Low-Light-Level Performance Analysis for Charge-Coupled Device TV Cameras,* 2
Kedem, Dan, *Methods and New Approaches to the Calculation of Physiological Parameters by Videodensitometry,* 118
Kedem, Drora, *Methods and New Approaches to the Calculation of Physiological Parameters by Videodensitometry,* 118
Kervitsky, J., *Low Light Level Imaging Devices for the Middle Ultraviolet,* 95
Lindstrom, D. P., *Methods and New Approaches to the Calculation of Physiological Parameters by Videodensitometry,* 118
Lynds, C. R., *Astronomical Applications of Charge Injection Devices,* 65
Morris, Clive W., *Isocon Imaging for X-Ray Diagnostics,* 108
Muehllehner, Gerd, *Image Intensifier Scintillation Cameras for Nuclear Medicine Applications,* 113
Nelson, J. H., *Methods and New Approaches to the Calculation of Physiological Parameters by Videodensitometry,* 118
Nelson, R. E., *Astronomical Applications of Charge Injection Devices,* 65
Nudelman, S., *High-Resolution Low-Light-Level Video Systems for Diagnostic Radiology,* 102
Opal, C., *Low Light Level Imaging Devices for the Middle Ultraviolet,* 95
Price, R. R., *Methods and New Approaches to the Calculation of Physiological Parameters by Videodensitometry,* 118
Rhea, T. C., Jr., *Methods and New Approaches to the Calculation of Physiological Parameters by Videodensitometry,* 118
Ricci, John L., *Isocon Imaging for X-Ray Diagnostics,* 108
Roehrig, H., *High-Resolution Low-Light-Level Video Systems for Diagnostic Radiology,* 102
———, *Results of Ratio Temperature Thermography,* 126
Sashin, Donald, *Isocon Imaging for X-Ray Diagnostics,* 108
Schuette, William H., *Application of Low Light Level Video Techniques to Biomedical Research,* 155
Shields, Herbert J., *Use of Night Vision Systems by the Land Manager,* 48
Smith, C. W., *Methods and New Approaches to the Calculation of Physiological Parameters by Videodensitometry,* 118
Sobieski, Stanley, *Intensified Charge Coupled Devices for Ultra Low Light Level Imaging,* 73
Stateham, Raymond M., *Detection of Mine Hazards with Infrared Imagers,* 34
Stupp, E. H., *Pyroelectric Vidicon Thermal Imager,* 23
Tegethoff, Jon, *General Application of Microchannel Image Inverters,* 28
Van den Berg, Alan J., *Infrared Is Not a Panacea—Use Discretion,* 41
Vescelus, Fred E., *A Large Imaging Array CCD Program,* 60
Way, Scott P., *Radiometric FLIR for Thermography,* 131
Whitehouse, Willard C., *Application of Low Light Level Video Techniques to Biomedical Research,* 155
Williams, Jack T., *Test Results on Intensified Charge Coupled Devices,* 78
Wolfe, W. L., *Results of Ratio Temperature Thermography,* 126

Subject Index

Aid as a Consumer Product, A Night Vision, 137
Analysis for Charge-Coupled Device TV Cameras, Low-Light-Level Performance, 2
Application of Low Light Level Video Techniques to Biomedical Research, 155
Application of Microchannel Image Inverters, General, 28
Applications, Image Intensifier Scintillation Cameras for Nuclear Medicine, 113
Applications of Charge Injection Devices, Astronomical, 65
Applications of Image Intensification Devices, Unusual, 55
Approaches to the Calculation of Physiological Parameters by Videodensitometry, Methods and New, 118
Array CCD Program, A Large Imaging, 60
Array Photometer, A Photon Counting, 83
Astronomical Applications of Charge Injection Devices, 65

Beam Scan vs Solid-State Readout, Comparison of TV Imagers for Use in Low-Light-Level Imaging by Electron, 14
Biological Surfaces—Excitation Source Brightness Requirements, Photoelectron Microscopy of, 143
Biomedical Research, Application of Low Light Level Video Techniques to, 155
Bombarded Mode, Operation of CCDs in the Electron, 10
Brightness Requirements, Photoelectron Microscopy of Biological Surfaces—Excitation Source, 143

Calculation of Physiological Parameters by Videodensitometry, Methods and New Approaches to the, 118
Cameras for Nuclear Medicine Applications, Image Intensifier Scintillation, 113
Cameras, Low-Light-Level Performance Analysis for Charge-Coupled Device TV, 2
CCD Program, A Large Imaging Array, 60
CCDs in the Electron Bombarded Mode, Operation of, 10
Charge-Coupled Device TV Cameras, Low-Light-Level Performance Analysis for, 2
Charge Coupled Devices for Ultra Low Light Level Imaging, Intensified, 73
Charge Coupled Devices, Test Results on Intensified, 78
Charge Injection Devices, Astronomical Applications of, 65
Comparison of TV Imagers for Use in Low-Light-Level Imaging by Electron Beam Scan vs Solid-State Readout, 14
Consumer Product, A Night Vision Aid as a, 137
Counting Array Photometer, A Photon, 83

Detection of Mine Hazards with Infrared Imagers, 34
Devices, Astronomical Applications of Charge Injection, 65
Devices for Middle Ultraviolet, Low Light Level Imaging, 95
Devices for Ultra Low Light Level Imaging, Intensified Charge Coupled, 73
Devices, Test Results on Intensified Charge Coupled, 78
Devices, Unusual Applications of Image Intensification, 55
Diagnostic Radiology, High-Resolution Low-Light-Level Video Systems for, 102
Diagnostics, Isocon Imaging for X-Ray, 108
Discretion, Infrared Is Not a Panacea—Use, 41

Electron Beam Scan vs Solid-State Readout, Comparison of TV Imagers for Use in Low-Light-Level Imaging by, 14
Electron Bombarded Mode, Operation of CCDs in the, 10

Excitation Source Brightness Requirements, Photoelectron Microscopy of Biological Surfaces–, 143

FLIR for Thermography, Radiometric, 131

General Application of Microchannel Image Inverters, 28

Hazards with Infrared Imagers, Detection of Mine, 34
High-Resolution Low-Light-Level Video Systems for Diagnostic Radiology, 102

Image Intensification Devices, Unusual Applications of, 55
Image Intensifier Scintillation Cameras for Nuclear Medicine Applications, 113
Image Inverters, General Application of Microchannel, 28
Imager, Pyroelectric Vidicon Thermal, 23
Imagers, Detection of Mine Hazards with Infrared, 34
Imagers for Use in Low-Light-Level Imaging by Electron Beam Scan vs Solid-State Readout, Comparison of TV, 14
Imaging Array CCD Program, A Large, 60
Imaging by Electron Beam Scan vs Solid-State Readout, Comparison of TV Imagers for Use in Low-Light-Level, 14
Imaging Devices for the Middle Ultraviolet, Low Light Level, 95
Imaging for X-Ray Diagnostics, Isocon, 108
Imaging, Intensified Charge Coupled Devices for Ultra Low Light Level, 73
Infrared Imagers, Detection of Mine Hazards with, 34
Infrared Is Not a Panacea–Use Discretion, 41
Injection Devices, Astronomical Applications of Charge, 65
Intensification Devices, Unusual Applications of Image, 55
Intensified Charge Coupled Devices for Ultra Low Light Level Imaging, 73
Intensified Charge Coupled Devices, Test Results on, 78
Intensifier Scintillation Cameras for Nuclear Medicine Applications, Image, 113
Inverters, General Application of Microchannel Image, 28
Isocon Imaging for X-Ray Diagnostics, 108

Land Manager, Use of Night Vision Systems by the, 48
Large Imaging Array CCD Program, 60
Low-Light-Level Imaging by Electron Beam Scan vs Solid-State Readout, Comparison of TV Imagers for Use in, 14
Low Light Level Imaging Devices for the Middle Ultraviolet, 95
Low Light Level Imaging, Intensified Charge Coupled Devices for Ultra, 73
Low-Light-Level Performance Analysis for Charge-Coupled Device TV Cameras, 2
Low-Light-Level Video Systems for Diagnostic Radiology, High-Resolution, 102
Low Light Level Video Techniques to Biomedical Research, Application of, 155

Manager, Use of Night Vision Systems by the Land, 48
Medicine Applications, Image Intensifier Scintillation Cameras for Nuclear, 113
Methods and New Approaches to the Calculation of Physiological Parameters by Videodensitometry, 118
Microchannel Image Inverters, General Application of, 28
Microscopy of Biological Surfaces–Excitation Source Brightness Requirements, Photoelectron, 143
Middle Ultraviolet, Low Light Level Imaging Devices for the, 95
Mine Hazards with Infrared Imagers, Detection of, 34
Mode, Operation of CCDs in the Electron Bombarded, 10

New Approaches to the Calculation of Physiological Parameters by Videodensitometry, Methods and, 118
Night Vision Aid as a Consumer Product, 137
Night Vision Systems by the Land Manager, Use of, 48
Nuclear Medicine Applications, Image Intensifier Scintillation Cameras for, 113

Operation of CCDs in the Electron Bombarded Mode, 10

Panacea–Use Discretion, Infrared Is Not a, 41
Parameters by Videodensitometry, Methods and New Approaches to the Calculation of Physiological, 118
Performance Analysis for Charge-Coupled Device TV Cameras, Low-Light-Level, 2
Photoelectron Microscopy of Biological Surfaces–Excitation Source Brightness Requirements, 143

Photometer, a Photon Counting Array, 83
Photon Counting Array Photometer, 83
Physiological Parameters by Videodensitometry, Methods and New Approaches to the Calculation of, 118
Product, A Night Vision Aid as a Consumer, 137
Program, A Large Imaging Array CCD, 60
Pyroelectric Vidicon Thermal Imager, 23

Radiology, High-Resolution Low-Light-Level Video Systems for Diagnostic, 102
Radiometric FLIR for Thermography, 131
Ratio Temperature Thermography, Results of, 126
Readout, Comparison of TV Imagers for Use in Low-Light-Level Imaging by Electron Beam Scan vs, 14
Requirements, Photoelectron Microscopy of Biological Surfaces–Excitation Source Brightness, 143
Research, Application of Low Light Level Video Techniques to Biomedical, 155
Resolution Low-Light-Level Video Systems for Diagnostic Radiology, High-, 102
Results of Ratio Temperature Thermography, 126
Results on Intensified Charge Coupled Devices, Test, 78

Scan vs Solid-State Readout, Comparison of TV Imagers for Use in Low-Light-Level Imaging by Electron Beam, 14
Scintillation Cameras for Nuclear Medicine Applications, Image Intensifier, 113
Security Systems, The Technology behind X-Ray, 44
Solid-State Readout, Comparison of TV Imagers for Use in Low-Light-Level Imaging by Electron Beam Scan vs, 14
Source Brightness Requirements, Photoelectron Microscopy of Biological Surfaces–Excitation, 143
Surfaces–Excitation Source Brightness Requirements, Photoelectron Microscopy of Biological, 143
Systems by the Land Manager, Use of Night Vision, 48
Systems for Diagnostic Radiology, High-Resolution Low-Light-Level Video, 102
Systems, Technology behind X-Ray Security, 44

Technology behind X-Ray Security Systems, 44
Temperature Thermography, Results of Ratio, 126
Test Results on Intensified Charge Coupled Devices, 78
Thermal Imager, Pyroelectric Vidicon, 23
Thermography, Radiometric FLIR for, 131
Thermography, Results of Ratio Temperature, 126
TV Cameras, Low-Light-Level Performance Analysis for Charge-Coupled Device, 2
TV Imagers for Use in Low-Light-Level Imaging by Electron Beam Scan vs Solid-State Readout, Comparison of, 14

Ultra Low Light Level Imaging, Intensified Charge Coupled Devices for, 73
Ultraviolet, Low Light Level Imaging Devices for the Middle, 95
Unusual Applications of Image Intensification Devices, 55
Use Discretion, Infrared Is Not a Panacea–, 41
Use in Low-Light-Level Imaging by Electron Beam Scan vs Solid-State Readout, Comparison of TV Imagers for, 14
Use of Night Vision Systems by the Land Manager, 48

Video Systems for Diagnostic Radiology, High-Resolution Low-Light-Level, 102
Video Techniques to Biomedical Research, Application of Low Light Level, 155
Videodensitometry, Methods and New Approaches to the Calculation of Physiological Parameters by, 118
Vidicon Thermal Imager, Pyroelectric, 23
Vision Aid as a Consumer Product, A Night, 137
Vision Systems by the Land Manager, Use of Night, 48

X-Ray Diagnostics, Isocon Imaging for, 108
X-Ray Security Systems, The Technology behind, 44